# Engineering Mathematics in Ship Design

# Engineering Mathematics in Ship Design

Special Issue Editors

**Cristiano Fragassa**
**Elizaldo Domingues Dos Santos**
**Nenad Djordjevic**

MDPI • Basel • Beijing • Wuhan • Barcelona • Belgrade

*Special Issue Editors*

Cristiano Fragassa
University of Bologna
Italy

Elizaldo Domingues Dos Santos
Universidade Federal do Rio
Grande—FURG
Brasil

Nenad Djordjevic
Brunel University London
UK

*Editorial Office*
MDPI
St. Alban-Anlage 66
4052 Basel, Switzerland

This is a reprint of articles from the Special Issue published online in the open access journal *Journal of Marine Science and Engineering* (ISSN 2077-1312) from 2018 to 2019 (available at: https://www.mdpi.com/journal/jmse/special_issues/eng_math).

For citation purposes, cite each article independently as indicated on the article page online and as indicated below:

LastName, A.A.; LastName, B.B.; LastName, C.C. Article Title. *Journal Name* **Year**, *Article Number*, Page Range.

**ISBN 978-3-03921-804-2 (Pbk)**
**ISBN 978-3-03921-805-9 (PDF)**

© 2019 by the authors. Articles in this book are Open Access and distributed under the Creative Commons Attribution (CC BY) license, which allows users to download, copy and build upon published articles, as long as the author and publisher are properly credited, which ensures maximum dissemination and a wider impact of our publications.

The book as a whole is distributed by MDPI under the terms and conditions of the Creative Commons license CC BY-NC-ND.

# Contents

About the Special Issue Editors . . . . . . . . . . . . . . . . . . . . . . . . . . . . . . . . . . . . . . vii

Cristiano Fragassa, Elizaldo Domingues dos Santos and Felipe Vannucchi de Camargo
Use of Engineering Mathematics for Ship Design
Reprinted from: *J. Mar. Sci. Eng.* **2019**, *7*, 370, doi:10.3390/jmse7100370 . . . . . . . . . . . . . . . 1

S.S. Kianejad, Jaesuk Lee, Yi Liu and Hossein Enshaei
Numerical Assessment of Roll Motion Characteristics and Damping Coefficient of a Ship
Reprinted from: *J. Mar. Sci. Eng.* **2018**, *6*, 101, doi:10.3390/jmse6030101 . . . . . . . . . . . . . . . 3

Rasul Niazmand Bilandi, Simone Mancini, Luigi Vitiello, Salvatore Miranda and Maria De Carlini
A Validation of Symmetric 2D + T Model Based on Single-Stepped Planing Hull Towing Tank Tests
Reprinted from: *J. Mar. Sci. Eng.* **2018**, *6*, 136, doi:10.3390/jmse6040136 . . . . . . . . . . . . . . . 22

Sakineh Fotouhi, Mohamad Fotouhi, Ana Pavlovic and Nenad Djordjevic
Investigating the Pre-Damaged PZT Sensors under Impact Traction
Reprinted from: *J. Mar. Sci. Eng.* **2018**, *6*, 142, doi:10.3390/jmse6040142 . . . . . . . . . . . . . . . 41

Riccardo Panciroli, Tiziano Pagliaroli and Giangiacomo Minak
On Air-Cavity Formation during Water Entry of Flexible Wedges
Reprinted from: *J. Mar. Sci. Eng.* **2018**, *6*, 155, doi:10.3390/jmse6040155 . . . . . . . . . . . . . . . 49

João Pedro T. P. de Queiroz, Marcelo L. Cunha, Ana Pavlovic, Luiz Alberto O. Rocha, Elizaldo D. dos Santos, Grégori da S. Troina and Liércio A. Isoldi
Geometric Evaluation of Stiffened Steel Plates Subjected to Transverse Loading for Naval and Offshore Applications
Reprinted from: *J. Mar. Sci. Eng.* **2019**, *7*, 7, doi:10.3390/jmse7010007 . . . . . . . . . . . . . . . 63

Felipe Vannucchi de Camargo
Survey on Experimental and Numerical Approaches to Model Underwater Explosions
Reprinted from: *J. Mar. Sci. Eng.* **2019**, *7*, 15, doi:10.3390/jmse7010015 . . . . . . . . . . . . . . . 75

Crístofer H. Marques, Jean D. Caprace and Alberto Martini
An Approach for Predicting the Specific Fuel Consumption of Dual-Fuel Two-Stroke Marine Engines
Reprinted from: *J. Mar. Sci. Eng.* **2019**, *7*, 20, doi:10.3390/jmse7020020 . . . . . . . . . . . . . . . 93

Youssef El Halal, Crístofer H. Marques, Luiz A. O. Rocha, Liércio A. Isoldi, Rafael de L. Lemos, Cristiano Fragassa and Elizaldo D. dos Santos
Numerical Study of Turbulent Air and Water Flows in a Nozzle Based on the Coanda Effect
Reprinted from: *J. Mar. Sci. Eng.* **2019**, *7*, 21, doi:10.3390/jmse7020021 . . . . . . . . . . . . . . . 105

Augusto Silva da Silva, Phelype Haron Oleinik, Eduardo de Paula Kirinus, Juliana Costi, Ricardo Cardoso Guimarães, Ana Pavlovic and Wiliam Correa Marques
Preliminary Study on the Contribution of External Forces to Ship Behavior
Reprinted from: *J. Mar. Sci. Eng.* **2019**, *7*, 72, doi:10.3390/jmse7030072 . . . . . . . . . . . . . . . 118

Jelena Šaković Jovanović, Cristiano Fragassa, Zdravko Krivokapić and Aleksandar Vujović
Environmental Management Systems and Balanced Scorecard: An Integrated Analysis of the Marine Transport
Reprinted from: *J. Mar. Sci. Eng.* **2019**, *7*, 119, doi:10.3390/jmse7040119 . . . . . . . . . . . . . . . **133**

# About the Special Issue Editors

**Cristiano Fragassa** is Adjunct Professor in Design of Machine at the Department of Industrial Engineering, University of Bologna, with over 20 years of experience in research and teaching. He is author of 200 publications, with more than half of them appearing internationally. His main fields of investigation have involved structural and safe design, composite materials, advanced process technologies. These topics have been approached both at the level of numerical modeling or experimental mechanics. Recently, his research directions have evolved to including sustainable design and circular economy in the case of ship design. He is also very active in terms of R&D or T&T, amassing a long list of internationally and nationally funded projects where he has been the coordinator or principal investigator.

**Elizaldo Domingues Dos Santos** is Associate Professor in Mechanical Engineering at the School of Engineering of Universidade Federal do Rio Grande (FURG) and involved in the graduate programs for ocean engineering and computational modeling as well as undergraduate courses in mechanical engineering. His main research themes have been concerned with convection heat transfer, turbulent flows, computational fluid dynamics, renewable energy, ocean engineering, and constructal design. He has a Scholarship in Research Productivity of CNPq—Brazil (Level 1D) and is author of 218 published journal papers, 85 of which appear in the Scopus database. He has also served as reviewer for several important journals which are indexed by Scopus and Web of Science databases. Moreover, he has served as a member in the advisory committee of numerous national and international funding agencies, and has served in several R&D nationally and internationally funded projects in engineering as coordinator or main investigator.

**Nenad Djordjevic** is Lecturer in Structural Integrity at National Structural Integrity Research Centre (NSIRC), Brunel University London, and Course Director of an MSc program, with 15 years of experience in research and teaching. He has been working on the development of linear and nonlinear numerical codes (FEM and SPH) for dynamic structural analysis, contributing to delivery of MSc programs in terms of program coordination, modular lecture delivery, and student project supervision, as well as presenting continuous professional development (CPD) courses. The main area of Nenad's research interests is in the development of constitutive models in the framework of thermodynamics and configurational mechanics, applicable to dynamic analysis of metals and composites and to the finite deformation problems. In particular, his research is oriented towards integrity analysis and simulation of a range of impact and crashworthiness problems in the aerospace, naval, and automotive industries, including bird strike, high velocity impact, and fluid–structure interaction. Another area of interest is the design and application of experimental techniques developed for the characterization of the dynamic behavior of materials. Nenad has been involved in several European programs, including the Horizon2020 project EXTREME, TEMPUS, FP6 and FP7 programs, and a number of industry sponsored projects developed in collaboration with Rolls Royce, AWE, McLaren F1, Lotus F1 (currently Renault), Catheram, Lockheed Martin, Office of Naval Research (USA), and numerous others. He is co-author of 19 papers published in eminent high-impact journals, and over 20 presentations at international conferences.

*Editorial*

# Use of Engineering Mathematics for Ship Design

Cristiano Fragassa [1,*], Elizaldo Domingues dos Santos [2] and Felipe Vannucchi de Camargo [1,3]

1. Department of Industrial Engineering, Alma Mater Studiorum Università di Bologna, Viale del Risorgimento 2, 40136 Bologna, Italy; felipe.vannucchi@unibo.it
2. School of Engineering, Universidade Federal do Rio Grande (FURG), Italia Av., km 8, 96211-090 Rio Grande, Brazil; elizaldosantos@furg.br
3. Post-Graduation Program in Mining, Metallurgical and Materials Engineering, Federal University of Rio Grande do Sul, Rua Osvaldo Aranha 99, 90035-190 Porto Alegre, Brazil
* Correspondence: cristiano.fragassa@unibo.it; Tel.: +39-347-697-4046

Received: 11 October 2019; Accepted: 17 October 2019; Published: 17 October 2019

---

With over that 70% of the Earth submerged by seas, continents separated by oceans, two thousand major islands scattered throughout the World, hundreds of thousands of kilometers of navigable rivers, maritime or fluvial transport surely represents one of the most important ways of moving people, goods, and wealth around the globe. This kind of transport has been widely used along the human history and development, allowing achievements not reachable in any other way in commercial, economic, technological, and even social aspects. As a consequence, the study of topics related to the ships design has always been a relevant aspect in every civilization and time. From the moment when the man has decided to face the sea on a boat up to the present day, a myriad of designers has posed the same fundamental question: 'are there knowledge or tools that can support the design of my boat in the way to make it as fast and safe as possible?'. This notable question was for who prepared the boat for the pharaoh about 3000 years before the Jesus Christ, as for who is designing the new catamaran for the next American's cup just now. However, fortunately, these centuries have allowed us to develop a profound knowledge of marine engineering, including a large range of methods and techniques useful for every boat designer.

Engineering mathematics, in particular, is a study field where mathematical methods and solving techniques are combined, regardless of the (analytical, numerical, or experimental) approach to offer a response to physical dilemma and to solve practical engineering problems.

In the present edition, this knowledge is employed to obtain several important recommendations about diversified fields of ship design, from the structural analysis in hulls up to methods that take into account the environmental sustainability in maritime transports.

Concerning the structural analysis and external forces over the vessels, this edition brings numerical studies about the influence of external forces (as the wind, inertial, and wave forces) over the trajectory of the ship and the effect of geometric configuration of stiffened plates under supported load in structures commonly found in hulls. It is also developed analytical methods based on single-stepped planning hulls for prediction of hydrodynamic forces over the structure of vessels. The influence of fluid structure interaction phenomena due to air cavity formation commonly found in naval applications and the employment of numerical methods for calculation of the roll motion in vessels are also investigated. Moreover, a survey on experimental and numerical approaches to model underwater explosions is explored, being a relevant study for adequate structural dimensioning of ships, especially for military vessels.

Investigations about propulsion systems are also presented. In this realm, the numerical study of propulsion devices based on Coanda effect are proposed and the evaluation of its main operational principle for thrust and maneuverability in marine applications is performed. Additionally, the estimated fuel consumption in dual-fuel two-stroke marine engines are analyzed, improving knowledge about selection of most efficient engines for naval applications.

To summarize, it is also performed an integrated analysis of the Environmental Management Systems and Balanced Scorecard for operation of marine transport taking into account environmental protection. We trust that this special edition will be helpful to improve comprehension of ship design and how the application of engineering mathematics can support the pursuit of knowledge in this research field.

**Conflicts of Interest:** The authors declare no conflict of interest.

© 2019 by the authors. Licensee MDPI, Basel, Switzerland. This article is an open access article distributed under the terms and conditions of the Creative Commons Attribution (CC BY) license (http://creativecommons.org/licenses/by/4.0/).

Article

# Numerical Assessment of Roll Motion Characteristics and Damping Coefficient of a Ship

S.S. Kianejad *, Jaesuk Lee, Yi Liu and Hossein Enshaei

Australian Maritime College, University of Tasmania, Launceston, TAS 7248, Australia; jaesukl@utas.edu.au (J.L.); yliu32@utas.edu.au (Y.L.); hossein.enshaei@utas.edu.au (H.E.)
* Correspondence: seyed.kianejadtejenaki@utas.edu.au; Tel.: +61-474-497-208

Received: 11 July 2018; Accepted: 27 August 2018; Published: 1 September 2018

**Abstract:** Accurate calculation of the roll damping moment at resonance condition is essential for roll motion prediction. Because at the resonance condition, the moment of inertia counteracts restoring moment and only the damping moment resists increase in the roll angle. There are various methods to calculate the roll damping moment which are based on potential flow theory. These methods have limitations to taking into account the viscous effects in estimating the roll motion, while, CFD as a numerical method is capable of considering the viscous effects. In this study, a CFD method based on a harmonic excited roll motion (HERM) technique is used to compute the roll motion and the roll damping moment of a containership's model in different conditions. The influence of excitation frequency, forward speed and degrees of freedom at beam-sea and oblique-sea realizations are considered in estimating the roll damping coefficients. The results are validated against model tests, where a good agreement is found.

**Keywords:** roll motion; roll damping; CFD; harmonic excitation

## 1. Introduction

Large roll motions in parametric roll and dead ship conditions are serious risks for the safety of a ship in rough sea conditions. To predict the roll motions accurately at resonance condition, estimation of the roll damping is essential. However, the accurate prediction of a ship roll damping is difficult, except by means of high cost experiments. Numerical approaches like CFD are an alternative option to estimate roll damping by considering the viscous effect.

In general, most of the roll damping calculation methods are based on potential flow theory and empirical method. The most common empirical method is Ikeda's method [1]. Though this method can be used quite well for conventional ships, the prediction results are sometimes conservative or underestimated for unconventional ships [2]. Roll damping is strongly nonlinear and is influenced by fluid viscosity and flow characteristics such as flow separation and vortex shedding. In theory, empirical or semi-empirical methods cannot take full consideration of different characteristics of a complex flow. Currently, vulnerability criteria for parametric roll and dead ship conditions are under development by the International Maritime Organization (IMO) as a second-generation intact stability criteria, in which roll damping coefficients are proposed, using Ikeda's simplified method. The calculation for traditional ships by Ikeda's simplified method can fit experimental data quite well at small roll angles. However, when the roll angle is large and out of the acceptable range of Ikeda's method, the accuracy of the damping coefficient is low.

In addition to Ikeda's simplified method, the correspondence group on Intact Stability regarding the second generation of intact stability criteria also proposed that the roll damping could be calculated by roll decay/forced roll test or CFD simulation. This suggestion can overcome the limitation of the model tests, which can predict roll damping very well but it is costly and time-consuming. Most of the experimental data is limited to a certain frequency range and particular geometry, which makes it impossible for the large-scale expansion of the application [3].

For the accurate calculation of roll damping, the influence of viscosity must be considered. CFD numerical simulations can consider flow characteristics and also reduce the cost of experiments. By improving CFD technology, it is possible to estimate damping coefficients precisely. Over recent years, numerous research projects based on CFD and experimental simulations have been conducted, for instance, Roddier [4] investigated two-dimensional simulations. The model was constrained to three degrees of freedom where it was free in roll, heave and sway motion. This numerical simulation used a random vortex method applied on a rectangular box in a beam wave condition. The results were validated against experimental data and showed that having bilge keels decreases roll resonance. They found that considering applied mechanical friction in the code can improve the accuracy of simulations. Na [5] carried out an investigation on a rectangular box with and without bilge keels to question harmonic force roll motion using experimental simulations. Various bilge keels' geometries were investigated with different width and angle to observe influences on the damping coefficient. As a result, it was found that bilge keels with larger lengths and horizontal orientation could improve the damping coefficient significantly.

Jung [6] used a particle imaging velocimetry (PIV) method to perform experimental simulations of a box to analyse vortex and turbulence generation in roll motion under beam waves conditions. It was found that fixing the box increases the intensity of turbulence due to intensification of relative velocity around the box. The flow around the corners was more turbulent because of separation. Yi-Hsiang [7] simulated the harmonic force roll motion of a floating production storage and offloading (FPSO) hull where the task was heavily involved with 2D CFD analysis. Bilge keels were attached in 45° inclination from the horizontal axis, which produced larger added moments of inertia and damping. The amplitude of added moments of inertia remained equal with horizontal or vertical bilge keels, while with the horizontal bilge keels it produced larger damping amplitude. Later, Kinnas [8] utilized the incompressible Navier-Stokes solver to analyse a 2D simulation of an FPSO hull in harmonic force roll motion with and without bilge keel appendages. The results showed that in inviscid flow, there is a linear relation between roll moment and roll angle amplitude. However, a non-linear variation exists between the roll moment and roll angle in viscous flow condition and it was observed that by adding the bilge keel, the nonlinearity could increase. Wilson [9] introduced numerical simulations based on unsteady Reynolds averaged Navier-Stokes (RANS) code to analyse the naval combatant's motions and wave patterns. The numerical simulations were conducted with and without bilge keels to investigate harmonic excited roll motion. The outcomes showed a good correlation with experimental data especially in case of a hull with bilge keels. However, the numerical approach had difficulties for simulating the free surface in a large roll angle.

Yu [10] utilized a 2D incompressible Navier-Stokes solver to investigate the roll damping of a rounded bilge box with and without bilge keels, as well as a sharp corner bilge box with and without a step at the keels. In that study, the exciting moment was subject to roll motion and it was observed that bilge keel increases the amplitude of damping and the relation between the roll moment and roll angle was nonlinear due to viscosity effects. Considering the roll damping of a rectangular barge, Bangun [11] performed a numerical simulation using a 2D incompressible Navier-Stokes solver in order to investigate roll motion. The model was examined with various conditions considering with and without bilge keel, different width sizes and angle of bilge keels. In total, 12 cases were examined. It was concluded that the barge with a smaller bilge keel angle from the horizontal axis produced larger roll damping. Thiagarajan [12] carried out experimental and numerical studies of FPSO in which the model was scaled 1:350. In numerical simulations, based on a free surface random vortex method (FSRVM), the model was forced to roll and results were in good agreement with the experimental data. Finally, it was concluded that the amplitude of damping is a function of roll angular velocity and width of bilge keels. An equation based on the relation between damping ratio and bilge keel width was proposed with some assumptions.

Avalos [13] performed a numerical simulation to investigate roll decay. The results gained from the numerical simulations were in an acceptable range which correlates well with experimental results. Through the process, the size of the vortex was a function of roll motion amplitude and width of the bilge keel. The roll decay technique is generally not a preferable method to estimate roll-damping coefficients in large roll motions especially with forward speed. In case of roll decay method, the water is initially at rest and roll dissipation occurs in a transient condition. Instead, damping obtained from harmonic excited roll motion (HERM) technique is based on steady state rolling motions, where the initial transient has already completed and the system is undergoing harmonic periodic motions. Therefore, the uncertainty of results is lower than the decay test where the roll angle magnitude will be decreases quickly over one cycle especially in case with forward speed. Blume [14] introduced a method to calculate roll damping coefficients effectively called the HERM technique, where this method excites the model in resonance frequency. However, this technique requires a longer time to determine the resonance frequency of the model. Another disadvantage of Blume's method is the dependency of the roll damping coefficient to maximum roll angle, metacentric height and heel angle, where each of them can be subjected to errors. Handschel [15] developed the HERM technique to estimate the damping coefficient in a range of frequencies that are very close to the resonance frequency. The technique considers the phase shift between the exciting moment and roll angles other than 90°. Begovic, Day [16] carried out CFD simulations using STAR CCM+ to calculate the roll damping of the DTMB 5415 trough roll decay technique for both intact and damage conditions. The results are compared against experimental measurement with reasonable accuracy. Mancini, Begovic Mancini, Begovic [17] conducted roll decay tests using numerical and experimental simulations to extract the roll damping coefficients. They considered the grid convergence index instead of correlation factor method to compute the uncertainty for the numerical simulations, because the solution was not close to asymptotic range. Zhou, Ning Zhou, Ning [18] conducted numerical and experimental simulations to estimate the roll damping of four different types of ships based on roll decay technique in zero forward speed. The results from experiments and numerical simulations were in good agreement. Somayajula and Falzarano [19] developed an advanced system of identification to compute frequency dependent roll damping from model test results in irregular waves. The results showed that the method can be used to predict a ship roll motion accurately compared to the potential flow method and empirical methods. Irkal and Nallayarasu Irkal, Nallayarasu [20] performed experimental and numerical simulations to compute the impact of bilge keels on roll damping. PIV method was used for the experiment to measure velocity field around the model during free oscillation tests. They found that the roll damping coefficient of the model without bilge keels is linear, however, with the bilge keels is strongly non-linear. Wassermann and Feder Wassermann, Feder [21] carried out model tests based on roll decay and HERM technique to calculate the roll damping of a container ship. They proposed various methods without additional filtering, curve fitting and offset manipulation of the recorded time series. They found that the HERM technique is more reliable in cases with higher forward speed and larger damping values. Oliva-Remola and Bulian Oliva-Remola, Bulian [22] conducted the HERM technique by shifting a mass harmonically inside the model in the lateral direction to generate excitation. The computed roll damping from HERM technique was smaller than roll decay tests for the same roll angle, because the model reached to a steady state rolling in HERM technique whereas the roll dissipation of roll decay tests occurs in a transient condition. They also proposed a 1 DOF mathematical model to predict the roll motion and calculate the roll damping. It was observed that tuning of dry roll inertia is critical to achieve good results, because the model was considered free in roll and sway motion.

There are a limited number of studies regarding the roll damping coefficients of an entire model and most studies considered a segment of the ship especially the middle section. The focus of numerical simulations was on 2D and overlooked the effect of other motions such as pitch and longitudinal turbulence in the case with forward speed. Therefore, the impact of different degrees of freedom (DOF) on the roll damping coefficient is unknown. In the present study, the numerical simulations of a whole containership model are conducted based on the HERM technique. The impact of forward speed, DOF and excitation frequency at beam sea and oblique sea conditions on roll motion characteristics and roll damping coefficients are investigated.

## 2. Theoretical Background

The following discussion involves the 'harmonic excited roll motion' HERM technique to determine required roll damping coefficients. Regarding Blume's experimental setup [14], the model possesses two masses at the centre of gravity that are rotating contrarily around the vertical axis. Specifically, one of the masses is rotating in a clockwise and the other is rotating in an anti-clockwise direction. The rotating mass shares the same frequency in the opposite direction to minimise the yaw motion. During contrary motion, the two masses meet at both sides of the model twice per rotation period, which imposes the maximum roll excitation moment. Various roll amplitudes can be achieved by setting different weights of the rotational masses. From the experiments [15], the amplitude of the roll exciting moment should ideally be equal to the restoring moment of the heel angle which refers to two masses of rotation on one side of the model. Originally, the equation of roll motion was formulated by using Newton's second law is balanced between the ships' motions and external moments as [23]:

$$[I_{44} + \delta I_{44}] \cdot \frac{d^2\varphi}{dt^2} + N_{44} \cdot \frac{d\varphi}{dt} + S_{44}\varphi = F_{E44}(t) \tag{1}$$

whereas,

| | |
|---|---|
| $I_{44} + \delta I_{44}$ | Mass and added mass moment of inertia coefficients |
| $N_{44}$ | Damping moment coefficient |
| $S_{44}$ | Restoring coefficient |
| $F_{E44}$ | Roll excitation moment |

The magnitude of damping moment in roll motion is generally less than the total moment of inertia and restoring moment. However, it still is essential to estimate the damping moment because inertia and restoring moments could be shifted in 180 degrees and counteract each other. As a result, only the damping moment limits the roll motion [24]. The exciting moment changes harmonically:

$$F_{E4}(t) = F_{E44}\sin(\omega t) \tag{2}$$

The induced roll angle and roll velocity are as below:

$$\varphi(t) = \varphi_a \sin(\omega t + \vartheta) \tag{3}$$

and

$$\frac{d\varphi}{dt} = \dot{\varphi}(t) = \omega \varphi_a \cos(\omega t + \vartheta) \tag{4}$$

The dissipated energy during a harmonic roll period is:

$$E_E = 4\int_0^{\varphi_a} N_{44}\dot{\varphi} d\varphi \tag{5}$$

Substituting the Equations (3) and (4) into Equation (5) with phase angle ($\vartheta = 0$) gives:

$$E_E = 4\int_0^{\frac{T}{4}} N_{44}\omega\varphi_a\cos(\omega t) \cdot \omega\varphi_a\cos(\omega t)dt = 4 \cdot N_{44}\omega^2\varphi_a^2 \cdot \int_0^{\frac{\pi}{2\omega}} \cos^2(\omega t)dt \tag{6}$$

Allowing the general solution of the integral gives:

$$\int \cos^2(xt)dt = \frac{2xt + \sin(2xt)}{4x} \tag{7}$$

The solution to define the dissipated damping energy over a cycle is:

$$E_E = 4 \cdot N_{44}\, \omega^2 \varphi_a{}^2 \cdot \left[\frac{2\omega t + \sin(2\omega t)}{4\omega}\right]_0^{\frac{\pi}{2\omega}} = \pi \cdot N_{44} \omega \varphi_a{}^2 \tag{8}$$

With respect to the roll moment, the roll angle is phase shifted by $\vartheta$. The following formula provides an estimation for the work done by the exciting moment in one roll period:

$$E_A = \int_0^T F_{E4}(t)\dot{\varphi}(t)dt = \int_0^{\frac{2\pi}{\omega}} F_{E44,a} \sin(\omega t)\omega \varphi_a \cos(\omega t + \vartheta)dt \tag{9}$$

Substituting Equation (4) into Equation (9) and after solving integral form, the formula gives:

$$E_A = F_{E44,a}\, \varphi_a \pi \sin \vartheta \tag{10}$$

The work done by the exciting moment and the dissipated energy over one roll period should relatively be the same. There are a couple of methods available to determine the roll damping coefficient, however, this paper focuses on the relationship of $E_A = E_E$ where the roll damping can be calculated by the following formula:

$$N_{44} = \frac{F_{E44} \sin \vartheta}{\omega \varphi_a} \tag{11}$$

The damping coefficient can now be established to a dimensionless form, which was recommended by the ITTC [25].

$$\hat{B}_{44} = \frac{N_{44}}{\rho \nabla B_{wl}^2} \sqrt{\frac{B_{wl}}{2g}} \tag{12}$$

## 3. Model Geometry

In this study, a model of a DTC post-Panamax container ship was adopted to carry out numerical simulations. A 3D sketch of the model is shown in Figure 1. The main particulars of the model scale and the full scale are presented in Table 1. The ship has a single screw propulsion with a five-blade arrangement and the ship's full appendages contain a rudder and bilge keels. The base profile of the rudder consists of an NACA 0018 foil. Both port and starboard sides have five bilge keel segments that are attached symmetrically. More details about the geometry of the model and ship can be found in [15].

**Figure 1.** A 3D geometry of the model and bilge keels set up at midsection.

Table 1. Main Dimensions of model and full-scale ship.

| Main Dimension | Full Scale (λ = 1.0) | Model Scale (λ = 59.467) |
|---|---|---|
| $L_{pp}$ [m] | 355.00 | 5.9697 |
| $L_{wl}$ [m] | 360.91 | 6.0691 |
| $B_{wl}$ [m] | 51.00 | 0.8576 |
| D [m] | 14.00 | 0.2354 |
| $C_B$ [-] | 0.6544 | 0.6544 |
| $\nabla$ [m³] | 165,868.5 | 0.7887 |
| $\overline{KM}$ [m] | 25.05 | 0.4213 |
| $\overline{GM}$ [m] | 1.37 | 0.023 |
| $\overline{KG}$ [m] | 23.68 | 0.3983 |
| $T_0$ [s] | 38.17 | 4.95 |
| $K_{XX}$ [m] | 20.25 | 0.340 |
| $K_{YY}$ [m] | 88.19 | 1.483 |
| $K_{ZZ}$ [m] | 88.49 | 1.488 |

## 4. Numerical Modelling

The right numerical set up is essential to achieve a successful numerical simulation. STAR CCM+ was used to conduct the study and this section provides details of the selected approach.

### 4.1. Governing Equations and Physics Modelling

The solver utilizes averaged continuity and momentum equations for incompressible flow in terms of tensor and Cartesian coordinates as follows [26]:

$$\frac{\partial(\rho \bar{u}_i)}{\partial x_i} = 0 \tag{13}$$

$$\frac{\partial(\rho \bar{u}_i)}{\partial t} + \frac{\partial}{\partial x_j}(\rho \bar{u}_i \bar{u}_j + \overline{\rho u'_i u'_j}) = -\frac{\partial \bar{p}}{\partial x_i} + \frac{\partial \bar{\tau}_{ij}}{\partial x_j} \tag{14}$$

$$\bar{\tau}_{ij} = \mu \left( \frac{\partial \bar{u}_i}{\partial x_j} + \frac{\partial \bar{u}_j}{\partial x_i} \right) \tag{15}$$

The flow directions are specified by $i$ and $j$ indices in x and y directions. Density and viscosity of the flow are represented by $\rho$ and $\mu$ respectively. $\bar{u}_i$ refers to the time-averaged velocity and $\bar{P}$ is the time-averaged pressure. The Reynolds stress tensor is illustrated by $\overline{\rho u'_i u'_j}$, while, $\bar{\tau}_{ij}$ is the mean viscous stress tensor. The selected solver is based on a finite volume approach method, which is a method for representing and evaluating partial differential equations in terms of algebraic equations. A predictor-corrector method is employed to form a relationship between continuity and momentum equations.

A turbulence model was employed due to the uncertainty of the stress tensor. In this paper, a realizable k-ε turbulence model was selected for the study. This selection decreases the simulation time in comparison to other turbulence methods such as SST and K-ω [27]. A simple multiphase approach called 'Volume of Fluid' (VOF) was employed to model the free surface. The mesh quality encompassing the free surface has the capability of solving interfaces between two phases. Hence, extra modelling was not necessary since the VOF method was selected. Simulations under VOF consider the same equation for a single phase or multiphase conditions, in which they also reflect the same velocity and pressure values. In order to capture the sharp interface among phases, the second order convection scheme was selected.

The solver utilizes the segregated flow model which becomes a useful tool to solve the governing equations in an uncoupled condition. Where convection terms were discretized by the second-order upwind scheme throughout the solution and the SIMMPLE algorithm was selected. To enhance the results, dynamic fluid body interaction (DFBI) method was used. This method is essential to predict the ship's behaviour in terms of seakeeping, because the model condition is like a ship in real sea condition [28]. A courant number (CFL) method was used to define the exact time step. In order to select the time step, the CFL method and recommended time step by ITTC [29] were considered. The CFL is a proportion of a physical time step to a mesh cell dimension per mesh flow velocity (Equation (16)) and it should be kept less than one for each cell to have numerical stability. However, 0.002 seconds ($T/2^{11}$ (T = excitation period)) was selected as a time step for the study which is quite smaller than both methods to capture accurately the roll motion and fluid-body interaction.

$$CFL = \frac{U \Delta t}{\Delta x} \qquad (16)$$

### 4.2. Meshing Structure

Overset mesh technique was used due to large motions involved in the simulations. The overset mesh does not need to any mesh modification after generating the initial mesh and the region can be transformed without remeshing, hence, it gives more flexibility and lower number of mesh [30]. This method involves two regions of overset and background. The overset region holds and surrounds the body and moves with the body while it is located inside the motionless background region. A linear interpolation method was used to taking into account the interaction between overset and background [28]. Because an overlap volumetric block was considered to generate the same cell size in both the background and overset regions at the vicinity of overset region to minimise the interpolation errors. To increase the accuracy, the overset region was refined in a more advanced mesh size and quality. This method allowed capturing boundary layer, flow separation during body motion, wave making and vortices around the body [31].

### 4.3. Mesh Generation

The mesh was generated according to the practical guidelines for ship CFD applications [32] and at least 40 cells per wavelength and 20 cells in the vertical direction for free surface were set. The trimmed mesher was utilised to generate a high-quality mesh and the prism layer was selected due to its capability to generate orthogonal prismatic mesh next to the body that can capture the velocity gradient and boundary layer. A surface remeshing option was used to produce a better quality of surface that can enhance volumetric mesh. For the final meshing stage, an automatic surface repair meshing tool was used to repair and purify any geometrical problems that were left over after surface remeshing. The volumetric control zones were produced around the body and in the free surface. The mesh refinement was performed mainly in these regions. Therefore, the number of cells was increased in those particular regions to capture complex flow characteristics. The cell size of the overset region and background were matched to prevent solution divergence. This was achieved by using an overlap volumetric block method. Figure 2 shows the mesh structure including the background and overset regions.

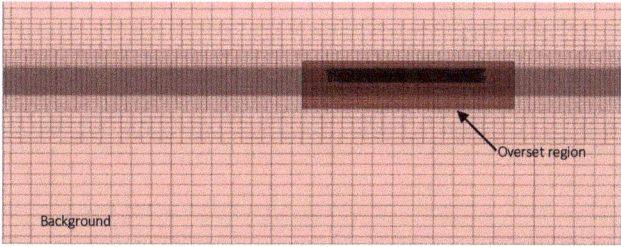

**Figure 2.** A 3D illustration of computational mesh.

## 4.4. Boundary and Initial Condition

To enhance the accuracy of results, suitable initial and boundary conditions were selected (Figure 3). The inlet boundary is set one $L_{pp}$ in front of the model and the outlet is located 3 $L_{pp}$ after the ship. The distances of left, right and bottom boundaries from the model are one $L_{pp}$ and 0.5 $L_{pp}$ is chosen for the top boundary [32]. The velocity inlet and pressure outlet boundary conditions were set in a way that the stream would flow past the model in head sea condition. The initial hydrostatic pressure was selected for the outlet boundary condition to prevent any backflow. The remaining boundaries including sides, top and bottom were all set as a velocity inlet in order to prevent a velocity gradient generated from the wall and flow interactions. Therefore, these boundary settings allow the velocity flow in all lateral boundaries to be directed towards the outlet boundary with negligible flow reflection.

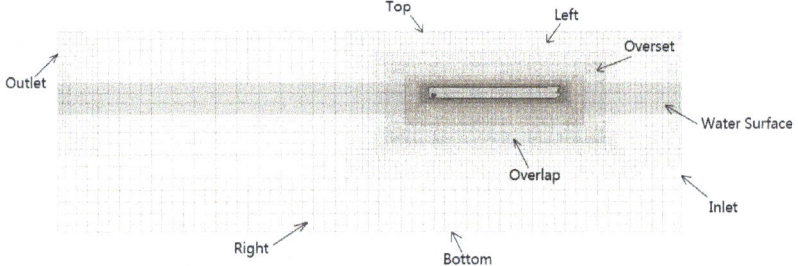

**Figure 3.** A 2D illustration of the overset and background regions with the applied boundary conditions.

## 4.5. Coordinate System

To simulate the model's motions, earth-fixed and body local coordinate systems were used. Initially, forces and moments on the model were calculated by analysing flow around the body. In the next step, the forces and moments were transferred into the local body coordinate where defined in the centre of gravity. The velocity and acceleration of the model were extracted from the motion's equations and converted to the earth-fixed coordinate system. This allows detecting the new location of the model.

## 5. Results and Discussion

### 5.1. Verification Analysis

To improve the reliability of the simulation results, it is necessary to specify the level of uncertainty. Based on the verification method presented by Stern [33], the numerical uncertainty $U_{SN}$, consists of uncertainty in iterative convergence $U_I$, grid-spacing $U_G$ and time-step $U_T$, which is formulated by the following equation:

$$U^2_{SN} = U^2_I + U^2_G + U^2_T \tag{17}$$

The uncertainty raised from $U_I$ and $U_T$ are negligible because the simulations are set in a calm water condition [34]. However, the grid-spacing uncertainty was investigated as major source of the uncertainty. Three different mesh configurations with 2.6, 3.6 and 4.5 million elements were created with a refinement ratio of $r_G = \sqrt{2}$ which was applied mainly on overset region [27,35] and details of the generated mesh are shown in Table 2. The model was excited by 5.5 Nm roll exciting moment based on the HERM technique at frequencies of 1.39 rad/s which is close to the natural frequency of the model. The drag calculation at a forward speed of 1.54 m/s was performed to select a proper mesh configuration that could precisely calculate pressure and shear forces. The increment for cells mainly focussed on the overset region to refine the quality of mesh. The accuracy of the simulation results against experimental measurements specifies which mesh configuration to be selected. The initial and boundary conditions all remained constant while the number of mesh cells varied. The grid uncertainty calculation for different mesh configurations was performed based on Richardson extrapolation [34]. The variation of simulation results for cases of coarse ($S_3$), medium ($S_2$) and fine ($S_1$) configurations are calculated as follows:

$$\varepsilon_{G32} = S_3 - S_2 \tag{18}$$

$$\varepsilon_{G21} = S_2 - S_1 \tag{19}$$

$$R_G = \varepsilon_{G21} / \varepsilon_{G32} \tag{20}$$

The numerical convergence ratio was calculated using Equation (20). Four typical conditions can be predicted for the convergence ratio: (i) monotonic convergence ($0 < R_G < 1$), (ii) oscillatory convergence ($R_G < 0$; $|R_G| < 1$), (iii) monotonic divergence ($R_G > 1$) and (iv) oscillatory divergence ($R_G < 0$; $|R_G| > 1$). For the cases (iii) and (iv) the numerical uncertainty cannot be computed. For the case (ii) uncertainty can be computed based on bounding error between upper limit $S_U$ and lower limit $S_L$ using Equation (21):

$$U_G = |\frac{1}{2}(S_U - S_L)| \tag{21}$$

In the case (i), the generalised Richardson extrapolation is adopted to compute the numerical uncertainty proposed by Stern, Wilson [36]. As the solutions were close to the asymptotic range, the correlation factor method was used to compute the numerical uncertainties. The maximum and minimum values of roll motion characteristics were taken into account to compute the uncertainty, because, the peak values are used to compute the roll damping.

Table 2. The number of mesh elements in different configurations.

|  | Background | Overset | Total |
|---|---|---|---|
| Fine (1) | 2,610,127 | 3,604,231 | 4,543,826 |
| Medium (2) | 1,725,314 | 2,449,946 | 3,326,086 |
| Coarse (3) | 884,813 | 1,154,285 | 1,217,740 |

The results of simulations are compared with experimental data [15] shown in Table 3 and the roll motion characteristics for a couple of cycles are shown in Figure 4. The numerical simulations were performed at 6DOF similar to experimental tests. Overall, the simulation results have produced larger values than experimental values. The verification study shows that the uncertainty value is small and about 5 for the worst condition. It was found that the 4.5 million mesh cells produce the most reliable results and could achieve the closest value to the experimental data. The maximum roll angle and drag have a 4.02% and 5.44% difference with experimental measurements, respectively. This numerical approach has the ability to simulate roll motion accurately. Hence, it was used to simulate roll motion characteristics in different conditions. In this study, a maximum of 3DOF (RHP) was considered for investigating dynamic stability according to the most popular equation methods, which have three degrees of freedom. Decreasing the number of degrees of freedom reduces the maximum roll angle because the moment of inertia and restoring moment do not fully counteract each other. Therefore, the computed roll damping moment, based on Handschel's method, is overestimated and using the proposed method in Section 5.4 can compute the roll damping coefficients more accurately.

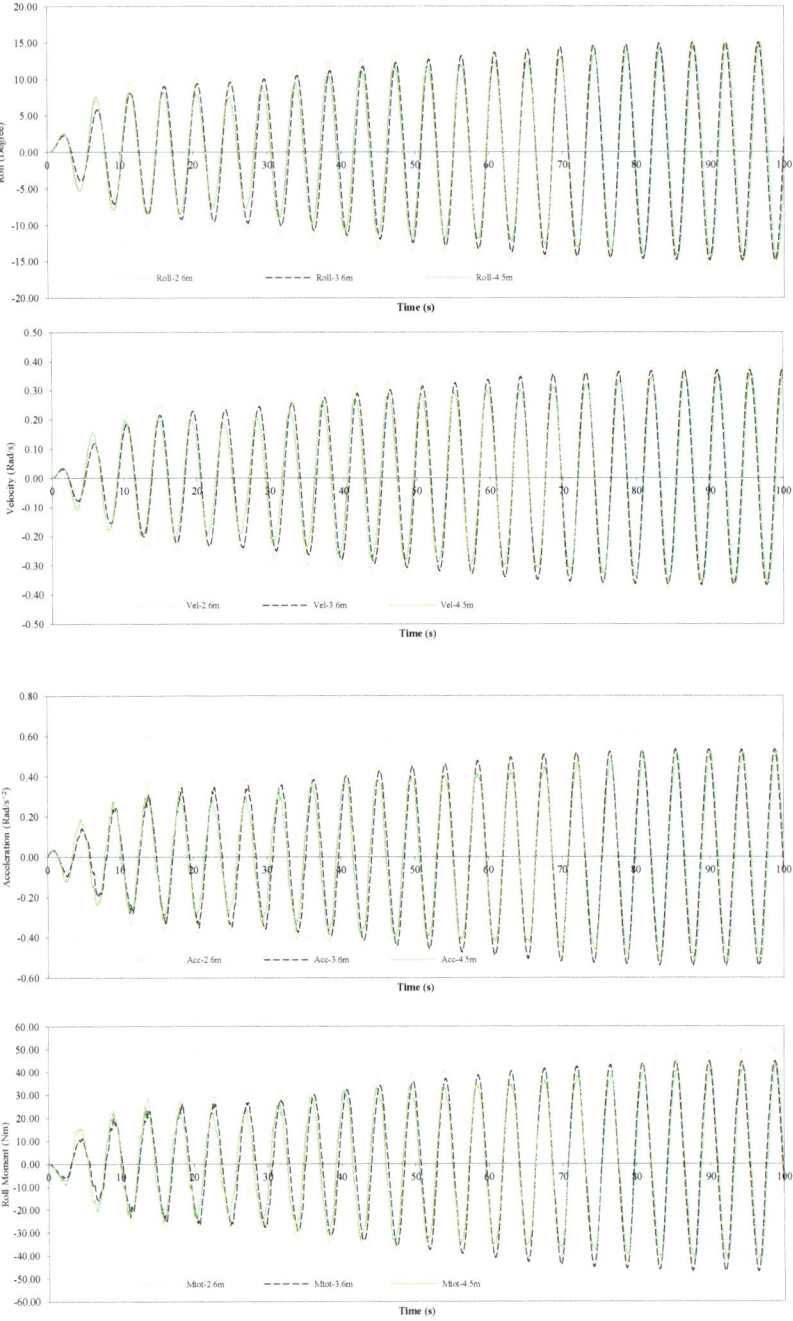

**Figure 4.** Influence of different mesh configurations on roll angle, angular roll velocity, angular roll acceleration and roll moment under 5.5 Nm roll exciting moment at frequency of 1.39 rad/s with zero forward speed and 6DOF.

Table 3. Verification Study of numerical results versus experiments (EFD).

| Amplitude | EFD | $S_1$ | $S_2$ | $S_3$ | $R_G$ | $\delta^*_{REG1}$ (%$S_1$) | $U_G$ (%$S_1$) |
|---|---|---|---|---|---|---|---|
| Roll angle (°) | 14.42 | 14.94 | 15.05 | 15.70 | 0.17 | 0.29 | 2.04 |
| Roll moment (Nm) | - | 43.8 | 44.5 | 50 | 0.13 | 0.47 | 4.42 |
| Angular acceleration (Rad/s$^2$) | - | 0.528 | 0.532 | 0.549 | 0.24 | 0.47 | 2.18 |
| Angular velocity (Rad/s) | - | 0.362 | 0.368 | 0.382 | 0.43 | 2.49 | 5.25 |
| Drag (N) | 26.46 | 27.9 | 28.14 | 30.5 | 0.10 | 0.19 | 2.36 |

## 5.2. Influence of Forward Speed and DOF

The roll damping coefficient cannot be computed directly from numerical simulations and requires time series analysis. Therefore, prior to discussing the damping coefficient, the influence of forward speed, degrees of freedom (DOF), different frequencies, beam sea and oblique sea conditions are studied. The model was excited based on the HERM technique at different conditions as presented in Table 4. In order to investigate the impact of DOF on roll motion characteristics, two conditions were considered (a) model free in just roll motion and restrain in other 5 DOF, (b) the model free in roll, heave and pitch (RHP) and restrain in surge, sway and yaw. Three different Froude numbers of 0, 0.1 and 0.19 were selected to investigate the impact of forward speed. In order to keep the results consistent, the physical simulation times were set to 20 s. The harmonic excited roll motion was generated from the beginning (zero degrees) and continued until the physical time of 20 s to demonstrate the motions of the model similar to real sea condition. This method enhances the accuracy of the motions for a better analysis. The changes of roll, angular velocity, angular acceleration and roll moment are shown in Figures 5–8. By observing the first two cycles of the plots, the model tends to absorb energy from the roll exciting moment in terms of the moment of inertia. Therefore, the roll motion characteristics increase gradually and remain constant in the following cycles. Decreasing the DOF to roll motion slightly decreases the amplitude of motion characteristics and the reduction is significant at higher Froude numbers. At zero forward speed, the roll angle difference between R and RHP conditions is 1.4 degrees, while the difference is 2.1 degrees at Froude number 0.19. Increasing the DOF increases the phase shift between the roll exciting moment and roll time trace (Figure 5). However, increasing the forward speed reduces the phase shift, because added dynamic pressure speeds up the rolling motion of the model. The angular velocity is a function of the roll time trace and a similar trend is observed. Increasing the forward speed and reducing the DOF reduce the amplitude of angular velocity (Figure 6). The model free in RHP at zero forward speed experiences higher angular acceleration and roll moment, whereas, the model free in just roll motion with the highest Froude number experiences lower angular acceleration and roll moment.

Table 4. Test conditions to study the impact of forward speed, DOF, excitation frequency, beam and oblique sea on roll motion and damping coefficients.

| Case No. | Fn. | Frequency (Rad/s) | DOF | Sea Condition | Roll Exciting Moment (Nm) | Pitch Exciting Moment (Nm) |
|---|---|---|---|---|---|---|
| 1 | 0 | 1.4 | R | Beam sea | 20 | - |
| 2 | 0 | 1.4 | RHP | Beam sea | 20 | - |
| 3 | 0.1 | 1.4 | R | Beam sea | 20 | - |
| 4 | 0.1 | 1.4 | RHP | Beam sea | 20 | - |
| 5 | 0.19 | 1.4 | R | Beam sea | 20 | - |
| 6 | 0.19 | 1.4 | RHP | Beam sea | 20 | - |
| 7 | 0 | 1.3 | RHP | Beam sea | 20 | - |
| 8 | 0 | 1.5 | RHP | Beam sea | 20 | - |
| 9 | 0 | 1.3 | RHP | Oblique sea | 20 | 10 |
| 10 | 0 | 1.4 | RHP | Oblique sea | 20 | 10 |
| 11 | 0 | 1.5 | RHP | Oblique sea | 20 | 10 |

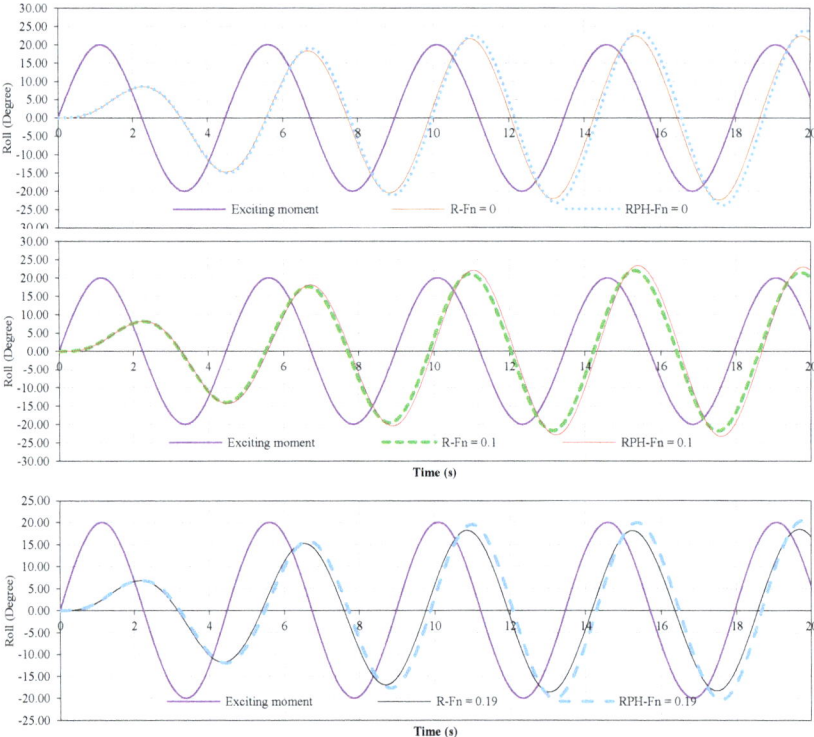

**Figure 5.** Roll angle time trace of the model under 20 Nm exciting moment at frequency of 1.4 rad/s, Fn = 0, Fn = 0.1, Fn = 0.19 and considering different degrees of freedom. Roll, heave and pitch motions are shown by R, H and P, respectively.

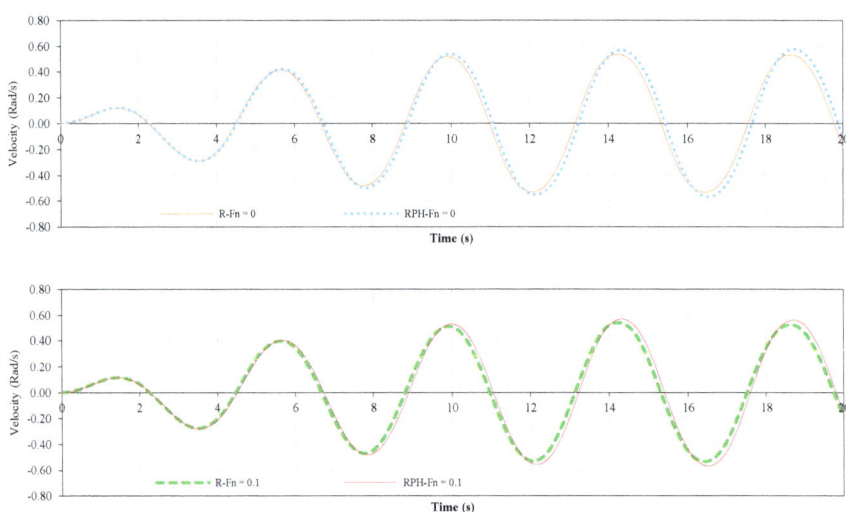

**Figure 6.** *Cont.*

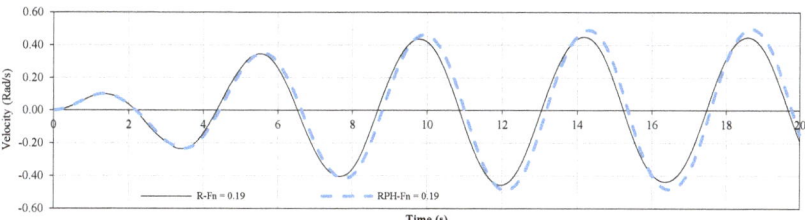

**Figure 6.** Angular roll velocity (Vel.) time trace of the model under 20 Nm exciting moment at frequency of 1.4 rad/s, Fn = 0, Fn = 0.1, Fn = 0.19 and considering different degrees of freedom. Roll, heave and pitch motions are shown by R, H and P, respectively.

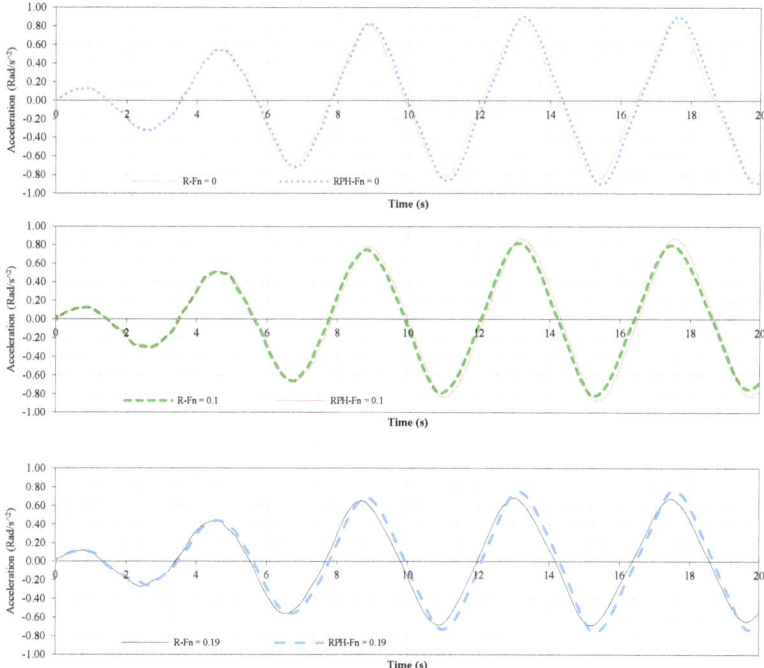

**Figure 7.** Angular roll acceleration (Acc.) time trace of the bare hull model under 20 Nm exciting moment at frequency of 1.4 rad/s, Fn = 0, Fn = 0.1, Fn = 0.19 and considering different degrees of freedom. Roll, heave and pitch motions are shown by R, H and P, respectively.

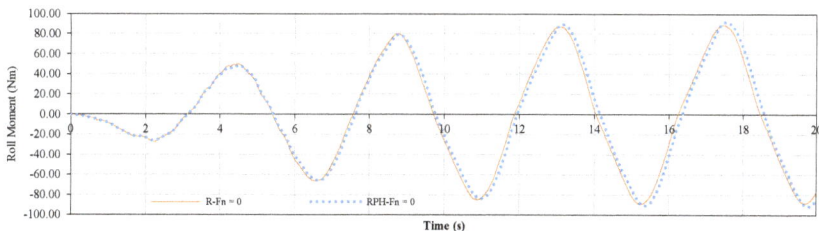

**Figure 8.** *Cont.*

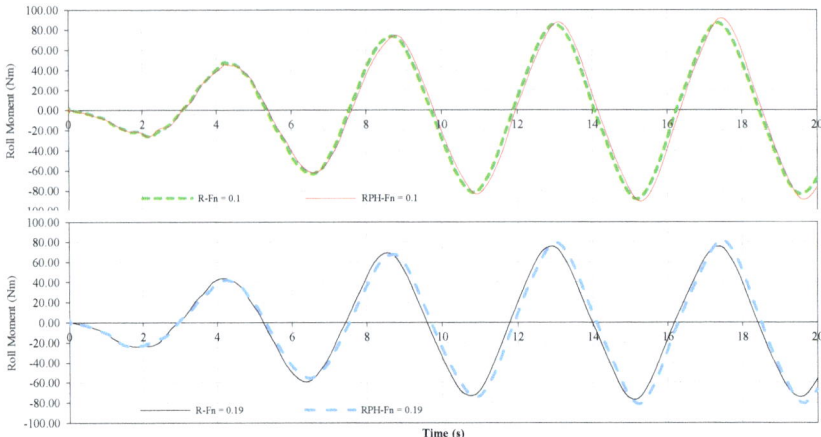

**Figure 8.** Total roll moment (Mtot) time trace of the model under 20 Nm exciting moment at frequency of 1.4 rad/s, Fn = 0, Fn = 0.1, Fn = 0.19 and considering different degrees of freedom. Roll, heave and pitch motions are shown by R, H and P, respectively.

### 5.3. Influence of Excitation Frequency and Heading

The model was excited in both beam sea and oblique sea conditions at different excitation frequencies including 1.3, 1.4 and 1.5 rad/s to investigate the effects of frequency and exciting moment direction on the roll motion characteristics. The variation of roll, angular velocity, angular acceleration, roll moment and pitch angle are shown in Figures 9–13. For the beam sea condition, 20 Nm roll exciting moment was applied on the model, while, for the oblique sea condition, 20 Nm and 10 Nm of roll and pitch exciting moment simultaneously were applied on the model respectively. The roll motion characteristics of the oblique sea conditions are slightly larger than the beam sea condition at various frequencies, although the total exciting moment in oblique condition is considerably larger. Because the model is more stable in longitudinal section and 10 Nm pitch exciting moment has negligible effects on pitch and roll motions. The model generates larger roll motion characteristics at a frequency of 1.4 rad/s which is close to the natural frequency of the model. The model achieves a 24-degree roll angle at 1.4 rad/s larger than 20.5 and 18 degrees for the frequency of 1.5 and 1.3 rad/s. At the resonance frequency, the induced moment of inertia and restoring moment have the same magnitude in the opposite direction hence counteracting each other. As a result, the damping moment which is not large resists the development of roll motion and the exciting moment imposes the model towards a larger roll angle. The angular velocity and roll moment of the model at a frequency of 1.4 rad/s is also larger than other frequencies, similar to the roll angle which is larger than the other two. However, the induced angular acceleration in the case of 1.5 rad/s is as large as the case of 1.4 rad/s. That is due to being located in the region dominated by acceleration where the magnitude of acceleration for the same amplitude of roll angle is larger compared to a lower range of frequency. The pitch angle changes with regard to the change of roll angle. The model experiences forward trim, because the wetted surface area and water pressure of the aft half of the model is larger than the forward half. The pitch angle at a frequency of 1.4 rad/s is larger due to a larger roll angle.

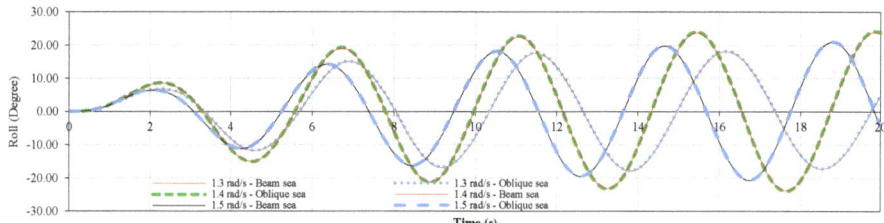

**Figure 9.** Roll angle time trace of the model under 20 Nm roll exciting moment at the frequencies of 1.3, 1.4 and 1.5 rad/s, Fn = 0 and considering beam sea and oblique sea conditions.

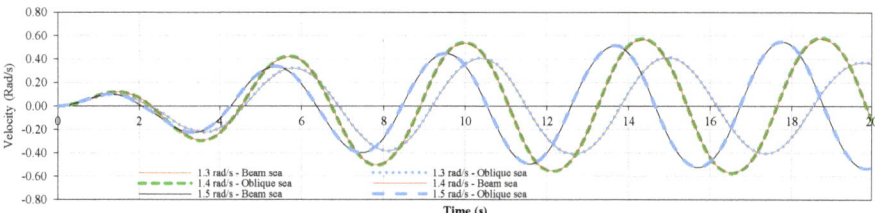

**Figure 10.** Angular roll velocity time trace of the model under 20 Nm roll exciting moment at the frequencies of 1.3, 1.4 and 1.5 rad/s, Fn = 0 and considering beam sea and oblique sea conditions.

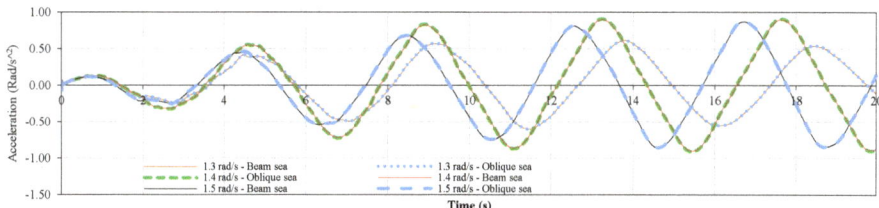

**Figure 11.** Angular roll acceleration time trace of the model under 20 Nm roll exciting moment at the frequencies of 1.3, 1.4 and 1.5 rad/s, Fn = 0 and considering beam sea and oblique sea conditions.

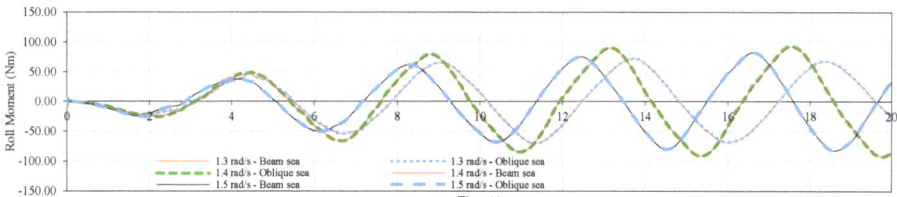

**Figure 12.** Total roll moment time trace of the model under 20 Nm roll exciting moment at the frequencies of 1.3, 1.4 and 1.5 rad/s, Fn = 0 and considering beam sea and oblique sea conditions.

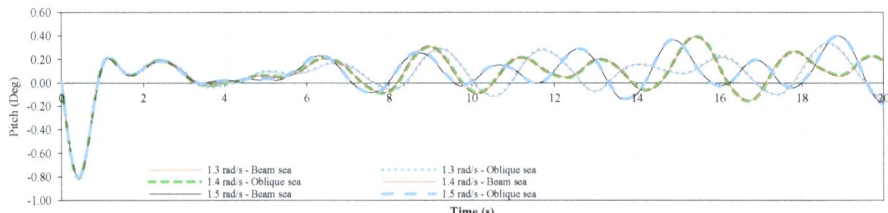

**Figure 13.** Pitch angle time trace of the model under 20 Nm roll exciting moment at the frequencies of 1.3, 1.4 and 1.5 rad/s, Fn = 0 and considering beam sea and oblique sea conditions.

## 5.4. Roll Damping Coefficient

The dimensionless damping coefficients of the model in different conditions are presented in Table 5. The damping coefficients were extracted from roll motion characteristics based on Equation (11) and were converted into a dimensionless number, which was recommended by the ITTC [25]. The model was free up to 3DOF and was excited at frequencies higher and lower than the natural frequency. Hence, the maximum roll motion characteristics were decreased compared to 6DOF at resonance conditions. That is because the total moment of inertia (IN) and restoring moment (RE) cannot counteract each other completely. The restoring moment was calculated by Autohydro software and the virtual moment of inertia was calculated based on [37]. The roll exciting moment compensates the difference between these two. Therefore, the amplitude of the real roll exciting moment decreases during a cycle.

**Table 5.** Non-dimensional roll damping of the model at different cases.

| Case No. | Maximum Roll Angle (Degrees) | Difference IN & RE (N) | Amplitude of Exciting Moment (N) | Phase Shift (Rad) | Dimensionless Damping Coefficient (CFD) | Dimensionless Damping Coefficient (EFD.) |
|---|---|---|---|---|---|---|
| 1 | 22.39 | 11.17 | 8.83 | 0.98 | 0.004816 | 0.00462 |
| 2 | 23.94 | 11.077 | 8.93 | 1.11 | 0.004907 | 0.00482 |
| 3 | 21.47 | 10.257 | 9.75 | 0.93 | 0.00535 | - |
| 4 | 22.98 | 10.11 | 9.89 | 1.02 | 0.00541 | - |
| 5 | 18.46 | 10.82 | 9.19 | 0.86 | 0.00553 | - |
| 6 | 20.48 | 10.57 | 9.44 | 1.01 | 0.00571 | - |
| 7 | 18.12 | 12.09 | 7.91 | 0.58 | 0.00377 | 0.00384 |
| 8 | 20.92 | 7.43 | 12.57 | 1.5 | 0.00822 | 0.00431 |
| 9 | 18.15 | 12.15 | 7.75 | 0.59 | 0.00376 | - |
| 10 | 24.18 | 11.14 | 8.86 | 1.11 | 0.00483 | - |
| 11 | 20.98 | 7.59 | 12.41 | 1.5 | 0.00809 | - |

It can be seen from cases 1 to 6 that reducing the DOF number at different Froude numbers reduces the maximum roll, whereas the damping coefficient relatively slightly increases (compared to the experimental values). The rudder angle was set at zero during the simulation to avoid any effect on extracting the damping coefficient. Increasing the forward speed increases the magnitude of the damping coefficient due to changes in pressure distribution around the model. However, the pressure variations are small and by reduction of the maximum roll angle, the damping coefficient increases. The damping coefficients of cases 1 and 2 are in acceptable agreement with existing results from the model tests [15]. Changing the excitation frequency changes the roll motion specifications and the damping coefficient. At frequencies of 1.3 rad/s (case 7) and 1.4 rad/s (case 2) the extracted damping coefficients are in good agreement compared to the experimental data, while for the frequency of 1.5 rad/s which is higher than the natural frequency the result is overestimated. Because, the energy conservation method to extract damping coefficient works well for frequencies equal or smaller than natural frequency of the model. For higher range of frequencies, the computed roll damping is overestimated because of the larger phase shift between roll exciting moment and roll angle, as well as smaller difference between virtual moment of inertia and restoring moment. This means the energy conservation method works at a range of frequency lower and close to the natural frequency. As discussed in the previous section, the roll motion characteristics in oblique sea condition are slightly larger than beam sea condition, while the computed damping coefficients of oblique sea condition are slightly smaller than the beam sea condition.

## 6. Conclusions

Numerical simulations based on the HERM technique were carried out to investigate the influence of excitation frequency, Froude number, DOF and the model's heading on roll motion and roll damping coefficients. If the model is free in just roll motion, it experiences smaller motion characteristics compared to the model free in coupled roll-pitch-heave motions. However, the reduction with a higher Froude number is more. Nevertheless, the model free in roll motion has relatively a larger roll damping coefficient at different forward speeds. The applied energy conservation method to calculate the damping moment is valid at a low range of frequencies as well as the natural frequency. The model under the same roll exciting moment in beam sea and oblique sea conditions generates similar roll motion and roll damping coefficients. The findings of this study improve the equation base methods to advance prediction of roll motion and to investigate the dynamic stability of a ship.

**Author Contributions:** S.S.K performed some of the simulations, the verification study, data analysis and the first draft of the manuscript. J.L. and Y.L. carried out the simulations and analysis of the results to extract the damping coefficients. H.E. helped to finalise the manuscript and supervised the research.

**Funding:** This research received no external funding.

**Conflicts of Interest:** The authors declare no conflict of interest.

## Nomenclature

| Symbol | Description (Unit) | Symbol | Description (Unit) |
|---|---|---|---|
| $F_{E44}(t)$ | Exciting moment (Nm) | $N_{44}$ | Damping coefficient (Nms/rad) |
| $F_{E44,a}$ | Amplitude of exciting moment (Nm) | $E_E$ | Dissipated energy (J) |
| $\omega$ | Frequency (Rad/s) | $E_A$ | Work done by roll exciting moment (J) |
| $\varphi$ | Roll angle (Degree) | $T$ | Period (s) |
| $\varphi_a$ | Amplitude of Roll angle (Degree) | $\rho$ | Density (Kg/m$^3$) |
| $v_1$ | Phase shift (Degrees) | $\nabla$ | Displacement (m$^3$) |
| $\dot{\varphi}$ | Angular velocity (Rad/s) | $B_{44}$ | Non-dimensional damping coefficient |

## References

1. Ikeda, Y.; Himeno, Y.; Tanaka, N. *A Prediction Method for Ship Roll Damping*; Department of Naval Architecture, University of Osaka Prefecture: Osaka, Japan, 1978.
2. Gu, M.; Lu, J.; Wang, T. Experimental and numerical study on stability under dead ship condition of a tumblehome hull. In Proceedings of the 13th International Ship Stability Workshop, Brest, France, 23–26 September 2013.
3. Bass, D.; Haddara, M. Nonlinear models of ship roll damping. *Int. Shipbuild. Prog.* **1988**, *35*, 5–24.
4. Roddier, D.; Liao, S.-W.; Yeung, R. Wave-induced motion of floating cylinders fitted with bilge keels. *Int. J. Offshore Polar Eng.* **2000**, *10*, 241–248.
5. Na, J.H.; Lee, W.C.; Shin, H.S.; Park, I.K. A design of bilge keels for harsh environment FPSOs. In Proceedings of the Twelfth International Offshore and Polar Engineering Conference, Kitakyushu, Japan, 26–31 May 2002.
6. Jung, K.H.; Chang, K.-A.; Huang, E.T. Two-dimensional flow characteristics of wave interactions with a free-rolling rectangular structure. *Ocean Eng.* **2005**, *32*, 1–20. [CrossRef]
7. Yi-Hsiang, Y.; Kinnas, S.A.; Vinayan, V.; Kacham, B.K. Modeling of flow around FPSO hull sections subject to roll motions: Effect of the separated flow around bilge keels. In Proceedings of the Fifteenth International Offshore and Polar Engineering Conference, Seoul, Korea, 19–24 June 2005.

8. Kinnas, S.A.; Yi-Hsiang, Y.; Vinayan, V. Prediction of flows around FPSO hull sections in roll using an unsteady Navier-Stokes solver. In Proceedings of the Sixteenth International Offshore and Polar Engineering Conference, San Francisco, CA, USA, 28 May–2 June 2006.
9. Wilson, R.V.; Carrica, P.M.; Stern, F. Unsteady RANS method for ship motions with application to roll for a surface combatant. *Comput. Fluids* **2006**, *35*, 501–524. [CrossRef]
10. Yu, Y.-H.; Kinnas, S.A. Roll response of various hull sectional shapes using a Navier-Stokes solver. *J. Offshore Polar Eng.* **2009**, *19*, 46–51.
11. Bangun, E.; Wang, C.; Utsunomiya, T. Hydrodynamic forces on a rolling barge with bilge keels. *Appl. Ocean Res.* **2010**, *32*, 219–232. [CrossRef]
12. Thiagarajan, K.P.; Braddock, E.C. Influence of bilge keel width on the roll damping of FPSO. *J. Offshore Polar Eng.* **2010**, *132*, 011303. [CrossRef]
13. Avalos, G.O.; Wanderley, J.B.; Fernandes, A.C.; Oliveira, A.C. Roll damping decay of a FPSO with bilge keel. *Ocean Eng.* **2014**, *87*, 111–120. [CrossRef]
14. Blume, P. Experimentelle Bestimmung von Koeffizienten der wirksamen Rolldämpfung und ihre Anwendung zur Abschätzung extremer Rollwinkel. *Schiffstechnik* **1979**, *26*, 3–23.
15. Handschel, S.; Abdel-Maksoud, M. Improvement of the Harmonic Excited Roll Motion Technique for Estimating Roll Damping. *Ship Technol. Res.* **2014**, *61*, 116–130. [CrossRef]
16. Begovic, E.; Day, A.H.; Incecik, A.; Mancini, S.; Pizzirusso, D. Roll damping assessment of intact and damaged ship by CFD and EFD methods. In Proceedings of the 12th international conference on the stability of ships and ocean vehicles, Glasgow, UK, 13–19 June 2015; pp. 14–19.
17. Mancini, S.; Begovic, E.; Day, A.H.; Incecik, A. Verification and validation of numerical modelling of DTMB 5415 roll decay. *Ocean Eng.* **2018**, *162*, 209–223. [CrossRef]
18. Zhou, Y.-h.; Ning, M.; Xun, S.; ZHANG, C. Direct calculation method of roll damping based on three-dimensional CFD approach. *J. Hydrodyn. Ser. B* **2015**, *27*, 176–186. [CrossRef]
19. Somayajula, A.; Falzarano, J. Application of advanced system identification technique to extract roll damping from model tests in order to accurately predict roll motions. *Appl. Ocean Res.* **2017**, *67*, 125–135. [CrossRef]
20. Irkal, M.A.; Nallayarasu, S.; Bhattacharyya, S. CFD approach to roll damping of ship with bilge keel with experimental validation. *Appl. Ocean Res.* **2016**, *55*, 1–17. [CrossRef]
21. Wassermann, S.; Feder, D.-F.; Abdel-Maksoud, M. Estimation of ship roll damping—A comparison of the decay and the harmonic excited roll motion technique for a post panamax container ship. *Ocean Eng.* **2016**, *120*, 371–382. [CrossRef]
22. Oliva-Remola, A.; Bulian, G.; Pérez-Rojas, L. Estimation of damping through internally excited roll tests. *Ocean Eng.* **2018**, *160*, 490–506. [CrossRef]
23. Enshaei, S.S.K. Quantifying Ship's Dynamic Stability through Numerical Investigation of Weight Distribution. In Proceedings of the 13th International Conference on the Stability of Ships and Ocean Vehicles (STAB), Kobe, Japan, 16–21 September 2018.
24. Kianejad, S.S.; Duffy, J.; Ansarifard, N.; Ranmuthugala, D. Ship Roll Damping Coefficient Prediction Using CFD. In Proceedings of the 32nd Symposium on Naval Hydrodynamics, Hamburge, Germany, 5–10 August 2018.
25. Reed, A.; Reed, A.M. 26th ITTC parametric roll benchmark study. In Proceedings of the 12th International Ship Stability Workshop, Washington, DC, USA, 2011.
26. Ferziger, J.H.; Peric, M.; Leonard, A. *Computational Methods for Fluid Dynamics*; American Institute of Physics (AIP): College Park, MD, USA, 1997.
27. Tezdogan, T.; Demirel, Y.K.; Kellett, P.; Khorasanchi, M.; Incecik, A.; Turan, O. Full-scale unsteady RANS CFD simulations of ship behaviour and performance in head seas due to slow steaming. *Ocean Eng.* **2015**, *97*, 186–206. [CrossRef]
28. CD-adapco, S. *STAR CCM+ User Guide Version 12.04*; CD-Adapco: New York, NY, USA, 2017.
29. ITTC. *Guidelines: Practical Guidelines for Ship CFD Applications*; Lyngby, Denmark, 2011; pp. 1–18.
30. Kianejad, S.S.; Duffy, J.; Ansarifard, N. Calculation of Restoring Moment in Ship roll motion through Numerical Simulation. In Proceedings of the 13th Int. Conference on the Stability of Ships and Ocean Vehicles (STAB), Kobe, Japan, 16–21 September 2018.
31. Field, P.L. *Comparison of RANS and Potential Flow Force Computations for the ONR Tumblehome Hullfrom in Vertical Plane Radiation and Diffraction Problems*; Virginia Tech: Blacksburg, VA, USA, 2013.

32. Procedures, I.-R. Practical Guidelines for Ship CFD Applications. In *Guidelines 2011*; 26th ITTC Specialist Committee on CFD in Marine Hydrodynamics; Lyngby, Denmark, 2011; pp. 1–18.
33. Stern, F.; Wilson, R.V.; Coleman, H.W.; Paterson, E.G. Comprehensive approach to verification and validation of CFD simulations-Part 1: Methodology and procedures. *Trans. Am. Soc. Mech. Eng. J. Fluids Eng.* **2001**, *123*, 793–802. [CrossRef]
34. Jin, Y.; Chai, S.; Duffy, J.; Chin, C.; Bose, N.; Templeton, C. RANS prediction of FLNG-LNG hydrodynamic interactions in steady current. *Appl. Ocean Res.* **2016**, *60*, 141–154. [CrossRef]
35. Tezdogan, T.; Incecik, A.; Turan, O. Full-scale unsteady RANS simulations of vertical ship motions in shallow water. *Ocean Eng.* **2016**, *123*, 131–145. [CrossRef]
36. Stern, F.; Wilson, R.; Shao, J. Quantitative V&V of CFD simulations and certification of CFD codes. *Int. J. Numer. Methods Fluids* **2006**, *50*, 1335–1355.
37. Kianejad, S.; Enshaei, H.; Ranmuthugala, D. Estimation of added mass moment of inertia in roll motion through numerical simulation. In Proceedings of the PACIFIC 2017 International Maritime Conference, Sydney, Australia, 3–5 October 2017; pp. 1–15.

© 2018 by the authors. Licensee MDPI, Basel, Switzerland. This article is an open access article distributed under the terms and conditions of the Creative Commons Attribution (CC BY) license (http://creativecommons.org/licenses/by/4.0/).

*Article*

# A Validation of Symmetric 2D + T Model Based on Single-Stepped Planing Hull Towing Tank Tests

Rasul Niazmand Bilandi [1,2], Simone Mancini [2,3,*], Luigi Vitiello [3], Salvatore Miranda [3] and Maria De Carlini [2]

1. Department of Engineering, Persian Gulf University, Bushehr 7516913817, Iran; rasool.niazmand@mehr.pgu.ac.ir
2. Eurisco Consulting Srls–R&D Company, 80059 Torre del Greco, Italy; maria.decarlini@euriscoconsulting.com
3. Department of Industrial Engineering (DII), University of Naples "Federico II", 80125 Naples, Italy; luigi.vitiello@unina.it (L.V.); salvatore.miranda@unina.it (S.M.)
* Correspondence: simone.mancini@unina.it; Tel.: +39-081-768-3308

Received: 8 October 2018; Accepted: 8 November 2018; Published: 12 November 2018

**Abstract:** In the current article, the hydrodynamic forces of single-stepped planing hulls were evaluated by an analytical method and compared against towing tank tests. Using the 2D + T theory, the pressure distribution over the wedge section entering the water and the normal forces acting on the 2D sections have been computed. By integrating the 2D sectional normal forces over the entire wetted length of the vessel, the lift force acting on it has been obtained. Using lift forces as well as the consequence pitch moment, the equilibrium condition for the single-stepped planing hull is found and then resistance, dynamic trim, and the wetted surface are computed. The obtained hydrodynamic results have been compared against the experimental data and it has been observed that the presented mathematical model has reasonable accuracy, in particular, up to Froude number 2.0. Furthermore, this mathematical model can be a useful and fast tool for the stepped hull designers in the early design stage in order to compare the different hull configurations. It should also be noted that the mathematical model has been developed in such a way that it has the potential to model the sweep-back step and transverse the vertical motions of single-stepped planing hulls in future studies.

**Keywords:** single-stepped planing hulls; symmetric 2D + T theory; hydrodynamic forces; towing tank tests

## 1. Introduction

Over time, researchers in naval architecture developed different methods to reduce the frictional resistance of planing hulls. Adding a transverse step in a high-speed monohull had been introduced as an appropriate method for reducing drag, for example, Step has a special geometry and enjoys improved hydrodynamic performance, i.e., resistance, dynamic stability, and seakeeping. Step can create a significant reduction of a dynamic wetted surface and of the dynamic trim angle during high-speed forward motion and thus achieve a reduction of resistance at high speed. There are four options for the hydrodynamic analysis of a stepped hull: the towing tank test [1–3], empirical method [4,5], analytical methods [6], and numerical simulation [7,8].

The performance prediction of a planing hull has long been used. For example, von Karman [9] and Wagner [10] modeled the wedge water entry as a planing section and computed the pressure distribution over the wedge surface. Wagner and von Karman initiated extensive research works in the Langley Memorial Aeronautical Laboratory and Davidson Laboratory in the United States. These studies were extended after the Second World War and continued until 1960. The authors conducted extensive sets of experiments and presented various empirical formulas for the prediction of planing hull characteristics. However, Savitsky [4] developed a mathematical model for the performance prediction of planing hulls. His model was able to predict lift, drag, wetted area, and the center of pressure of non-stepped planing hulls in calm waters. The Savitsky model was based on formulas derived from the aforementioned experiments.

The basis of the Savitsky method created an incentive for other researchers to expand the empirical formulas to calculate the lifting forces and modify the Savitsky method for stepped planing hulls. Svahn [11] first developed a mathematical model for the performance prediction of a stepped hull. His model can only simulate a one-stepped planing hull and uses Savitsky and Morabito's [12] formulas for separating flow from the step. However, Danielsson and Stromquist [13] mentioned that Savitsky and Morabito's [12] formulas cannot be implemented for a two-stepped hull because these relations have basically been derived for transom stern flow—not for separated flow from steps. Therefore, Dashtimanesh et al. [5] assumed a linear wake theory and presented a simplified mathematical model for the performance prediction of two-stepped planing hulls. The authors compared the obtained results with the experiments of Taunton et al. [1] and Lee et al. [2] and showed that their mathematical model has a good accuracy. Their model was based on Savitsky's formulas and regression relation for lift force which is limited by a special range for trim, wetted length, and the speed coefficient. Moreover, it was not possible to compute the pressure distribution over the hull length.

After these empirical studies, the researchers used the 2D + T or numerical method to look at the hydrodynamics of the planing hull or planing hull section in calm waters. The accuracy of the numerical method is high but has many complexities and cannot easily be used in the initial phases of design. The application of 2D + T theory dates back to the end of the 1970s, where Zarnick [14] utilized Wagner [10] and von Karman [9] theoretical equations and developed a mathematical model based for computation of planing hull behavior in waves. This method could be used for the performance prediction if the water surface is set to be at rest. The constant heave and pitch result correspond to the sinkage and dynamic trim of a vessel. Ghadimi et al. [15,16] extended Zarnick's method for the motion prediction of planing hulls in regular waves at 4 and 6 degrees of freedom, successively. Moreover, Ghadimi et al. [17–19] and Tavakoli et al. [20–23], developed several mathematical models for computation of roll motions in waves, roll motion, asymmetric, and yawed condition motions. All of these studies are performed using 2D + T theory and relate to non-stepped planing hulls. Thus, in this work, the 2D + T theory method is used and the hydrodynamic pressure over the wall of the 2D wedge is utilized to find the sectional hydrodynamic forces in the stepped planing hull, and the performance of the vessel is solved.

In this paper, the main aim is to develop a mathematical model for the simulation of single-stepped planing hull characteristics by using the 2D + T theory and linear wake assumptions. The basis of the present mathematical model is taken from the mathematical model developed by Niazmand Bilandi et al. [6]. The hull of a single-stepped planing hull has been divided into two parts; for each part, a water entry problem has been simulated. The hydrodynamic pressure on the 2D sections of the single-planing hull is predicted by the Algarin and Tascón [24] equations. The forces acting on each part of the boat are determined using the 2D + T theory. The main results, including the dynamic trim angle, wetted surface, and resistance, have been computed with the proposed method and have been compared against experimental data. In Section 2, a mathematical model and computer procedure are demonstrated; in Section 3, validation and results are presented. In particular, the model test and experimental details, a comparison between the towing tank tests and 2D + T method resulting in the term of resistance, wetted surface, and wetted length analysis are presented. Section 4 presents the conclusions.

## 2. Mathematical Model

The presented method is formulated in this section. The problem is defined and motion equations, as well as 2D + T theory, are discussed. The computation of hydrodynamic force acting on the hull has been fulfilled based on the pressure distribution on the wedge surface. To calculate the hydrodynamic pressure, the resulting forces and moments on a single-stepped planing hull, equations for the simulation of the water entry of wedges and estimation of a half-wetted beam are used, and a new mathematical model for the performance prediction of single-stepped planing hulls with a pressure-based approach was developed.

Further, it has been considered that the boat is moving forward with a constant speed of V and dynamic trim angle of $\theta$, demonstrated in Figure 1. The defined measure of dynamic trim ($\theta$) for the 2D + T method is relative to the keel line of the hull. For a constant deadrise (pure wedge) hull, the keel line is straight and parallel to all the buttock flow lines, so the geometric definition of trim is clear. The dynamic trim angle depends on the boat speed. Figure 1 includes the weight force ($\Delta$), the force derived from the hydrodynamic and hydrostatic pressure (F), the total frictional drag force (D), and the thrust force (T), derived from the various planing surfaces on the boat. Additionally, in Figure 1, two right-handed coordinate systems are adopted. The $G\xi\eta\zeta$ system is fixed on the body and located at CG. In this coordinate system, $\xi$ is parallel to the keel and positive forward, $\eta$ is positive in the direction of the starboard side, and $\zeta$ is positive downward. The $Oxyz$ system is moving with the system. The x-axis is parallel to the calm water and positive forward. The boat with deadrise angle $\beta$ has been fixed at the zero heel angle in the calm water, $\tau_1$ and $\tau_2$ are the local trim angles for each planing surface.

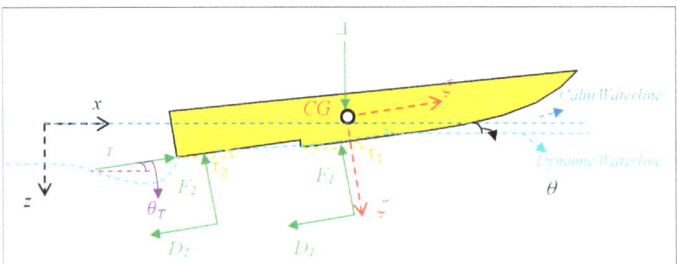

**Figure 1.** The problem definition and coordinate systems.

The equilibrium in the heave and pitch directions, as shown in Figure 1, can be calculated from the following equations.

$$0 = \sum_{i=1}^{2} (T_z + D_{z_i} + F_{z_i} + \Delta) \quad \text{Heave} \tag{1}$$

$$0 = \sum_{i=1}^{2} (T_\theta + D_{\theta_i} + F_{\theta_i}) \quad \text{Pitch} \tag{2}$$

To solve the problem, some assumptions are made as follows:

- The speed, V, was assumed to be constant for all two planing surfaces. In reality, the speed of the water would decrease aft of each step due to disturbances from the hull and turbulence. This would implicate that the lift from the middle and aft planing surface would be slightly exaggerated. By applying the effects of the transom and the steps, the forces will be calculated with a more accurate value.
- The planing surfaces are assumed to have triangular shapes.

- The wake profile is considered horizontal and parallel to the horizon from the separation at the step to where it reattaches on the next surface, contradicting Morabito's wake theory [12], Svahn [11], Dashtimanesh et al. [5], and Niazmand Bilandi et al. [6] suggested this simplification.
- The sweep-back of the steps is not included in the model.
- The local deadrise angle $\beta_L$ has been assumed to be 2 degrees for each planing surface. This value depends on the ventilation length and has effects on the trim and resistance of the vessel because it affects the lift coefficients.
- The local trim angle, $\tau$, has also been assumed to be 2 degrees. This value is measured using the slope of the planing surface in relation to the horizon, which has a straightforward relationship with step height.

In the first step, the total wetted length $L_{w_0}$, and the overall dynamic trim angle $\theta$, should be estimated. Therefore, the wetted length of the front planing surface can be calculated using Equation (3).

$$L_{w_1} = L_{w_0} - L_s \tag{3}$$

where, $L_s$ is the step position.

In the current article, an attempt has been made to develop a novel mathematical model based on the studies of Danielsson and Stromquist [13], Dashtimanesh et al. [5], and Niazmand Bilandi et al. [6]. As mentioned in literature reviews, there has been no direct measurement or empirical formula for wake profile beneath the stepped hulls. Therefore, Danielsson and Stromquist [13] observed that the linear wake profile (LWP) may be a good assumption for the flow separation from the steps. Therefore, the present study attempts to take into account this suggestion for a single-stepped planing hull. So, for the single-stepped planing hull, the ventilation length is calculated from the following equation.

$$L_{dry} = \frac{H_{step}}{\tan(\theta + \tau_1)} \tag{4}$$

where, $H_{step}$ is the steps height, and $\tau_1$ is local trim angle for the forward planing surface.

Subsequently, the wetted length is also calculated for the transom planing surface Equation (4).

$$L_{w_2} = L_s - L_{dry} \tag{5}$$

In this paper, for calculating various planing characteristics on the single-stepped planing hulls, the 2D + T theory has been formulated for each surface, individually. As shown in Figure 2, it has been assumed that the boat passed through an earth-fixed plane for each planing surface. In this regard, the vertical impact velocity and time needed to solve symmetric wedge water entry problem for each planing surface are calculated as follows.

$$w_i = v \sin(\theta + \tau_i); \ (i = 1, 2) \tag{6}$$

$$t_{p_i} = \frac{L_{w_i}}{v}; \ (i = 1, 2) \tag{7}$$

where $L_{w_i}$ is the wetted length of each planing surface.

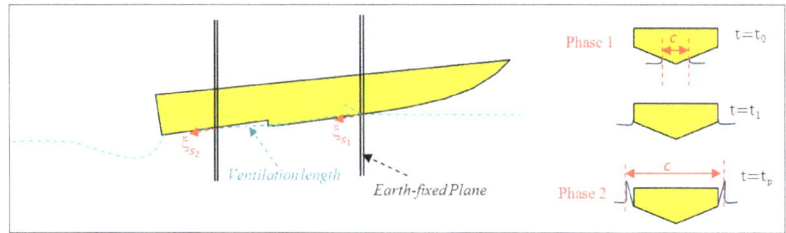

**Figure 2.** The 2D + T theory for single-stepped planing hull; it passes through a fixed two-dimensional observation plane for each planing surface (**left**); water entry problem for each planing surface from $t = t_0$ to $t_p$ (**right**).

To convert the time to the longitudinal position, Equation (8) is utilized for each planing surface.

$$\tilde{\zeta}_{s_i} = \frac{vt}{\cos(\theta + \tau_i)}; \ (i = 1, \ 2) \tag{8}$$

So, the longitudinal distance of the section from the intersection of the calm water and keel of each surface ($\tilde{\zeta}_{s_i}$) is computed. This position can be transformed to the body-fixed coordinate system by using

$$\tilde{\zeta}_1 = (L_{k_1} + L_{step_1} - \tilde{\zeta}_{s_1}) \tag{9}$$

$$\tilde{\zeta}_2 = (L_{k_2} - \tilde{\zeta}_{s_2}) \tag{10}$$

### 2.1. Two Dimensional Forces

The hydrodynamic pressure distribution over the surface of a symmetric wedge section has been calculated using the analytical solution of Wagner [10] as follows,

$$p_i = \rho \left( -\frac{w_i c_i \dot{c}_i}{\sqrt{c_i^2 - y_i^2}} - \frac{w_i^2}{2} \frac{y_i^2}{c_i^2 - y_i^2} \right); (i = 1, \ 2) \tag{11}$$

where $c$, $y$, and $\dot{c}$ are the half wetted beam, the horizontal distance from the keel and derivative $c$ with respect to time, respectively (Figure 2).

Generally, Equation (11) should be used to determine the pressure distribution over the surface of a symmetric wedge on each planing surface. To compute the hydrodynamic pressure, the proposed method by Algarın and Tascon [24] has been utilized. It should be noted that two different phases are considered in the computations which are related to the water depth location, as shown in Figure 2.

### 2.2. Phase 1—The Dry-Chine Condition

The first phase is related to the dry-chine condition in which chine will remain dry. Spray root position at each side of the wedge surface and its time derivative can be determined by Equations (12) and (13). The symmetry wedge section and spray root position are shown in Figure 2. Using of the Equations (11)–(13), the pressure distribution on both sides of the wedge can be calculated.

$$c_i = \frac{\pi}{2} w_i t \tan(\beta + \beta_{L_i}); \ i = 1, 2 \tag{12}$$

$$\dot{c}_i = \frac{\pi}{2} w_i \tan(\beta + \beta_{L_i}); \ i = 1, 2 \tag{13}$$

where $\beta$ is the deadrise angle and $\beta_{L_i}$ is the local deadrise angle. At very high speeds, the local deadrise angle will be almost parallel to the deadrise of the hull (Figure 3) so that the local deadrise angle will be in the magnitude of 2–4 degrees. This matches the values found by Svahn [11].

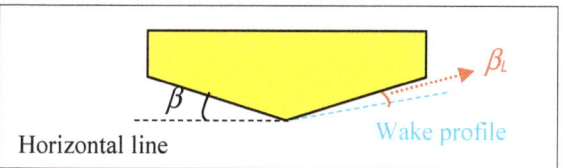

**Figure 3.** The deadrise angle of the stepped hull.

### 2.3. Phase 2—The Wet-Chine Condition

The second phase refers to the condition in which the chines of two sides become wet. The chine wetting time at both sides can be determined as follows,

$$t_{cw_i} = \frac{1}{\pi} \frac{B \tan(\beta_i + \beta_{Li})}{w_i}, i = 1, 2 \tag{14}$$

When the chines have been wetted, the idea by Algarın and Tascon [24] was used for the calculation of pressure distribution. So, after the chines are wetted, the mean half beam and its derivative, are given approximated by Equations (15) and (16).

$$c_i = \sqrt{\left(\frac{B}{2}\right)^2 + \left[\frac{3}{2}(w_i)\left(\frac{B}{2}\right)^2 (t - t_{cw_i})\right]^{2/3}}; i = 1, 2 \tag{15}$$

$$\dot{c}_i = \frac{w_i^2}{2} \frac{\left(\frac{B}{2}\right)^2}{c_i \sqrt{c_i^2 - \left(\frac{B}{2}\right)^2}}; i = 1, 2. \tag{16}$$

By integrating the hydrodynamic pressure over the wetted surface, the hydrodynamic forces acting on the wedge for each plane surface has been obtained as follows,

$$f_{HD_i}^V = \int_S p_i \cos(\beta_i + \beta_{L_i}) dl; i = 1, 2 \tag{17}$$

$$f_{HD_i}^H = \int_S p_i \sin(\beta_i + \beta_{L_i}) dl; i = 1, 2 \tag{18}$$

where, superscripts $V$ and $H$ refer to the vertical (normal force) and horizontal components of the hydrodynamic force, respectively, and $l$ is the distance from the wedge apex in the direction of the wedge wall (m).

The hydrostatic force of each section is determined by calculating the volume of the immersion for each planing surface (Figure 4), according to the following equation.

$$f_{B_i} = \rho g A_i; i = 1, 2 \tag{19}$$

When the chine is dry, $A_i$ is calculated as follows.

$$A_i = \frac{c_i^2}{\tan(\beta_i + \beta_{L_i})}; i = 1, 2 \tag{20}$$

When the chine is wet, $A_i$ is calculated as follows.

$$A_i = \frac{B^2}{2 \tan(\beta_i + \beta_{L_i})}; i = 1, 2 \tag{21}$$

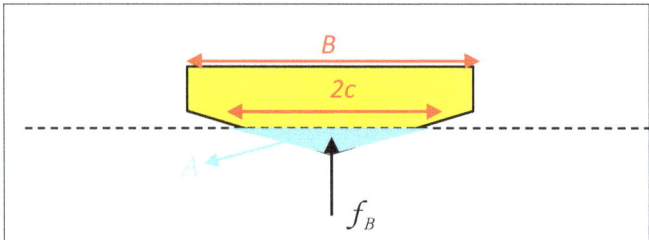

**Figure 4.** The hydrostatic force acting on the section.

## 2.4. Three Dimensional Forces

The 2D pressure forces have been determined during the water entry problem for each plane surface. These pressure forces have been extended over the wetted length of the boat and lead to the computation of 3D forces for each plane surface. By applying Garme's function [25] to the effects of the transom and the steps, the forces will be calculated with a more accurate value. This function is given as,

$$C_{tr_i} = \tanh(\frac{2.5}{a}(\xi^j - \xi_i));\ i = 1,\ 2 \tag{22}$$

where $j$ is the step position or transom position (Figure 5), $a = BFn_B a_{non}$ and $a_{non}$ is the non-dimensional longitudinal position (from the transom and step) in which the reductions appear. Garme [25] proposed that the anon be set to 0.34 for the planing range.

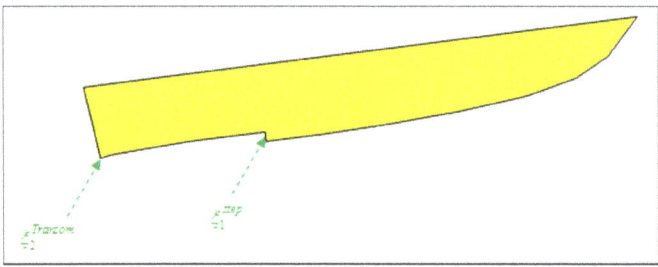

**Figure 5.** The steps and transom position.

The pressure force in the surge direction can be determined as follows,

$$F_{x_i} = -\int_{L_{w_i}} f^v_{HD_i} C_{tr_i}(\xi) \sin(\theta + \tau_i) d\xi;\ i = 1,\ 2 \tag{23}$$

Additionally, the total lift force in the heave direction has been determined by the summation of 3D forces for each plane section as follows,

$$F_{z_i} = -\int_{L_{w_i}} f^v_{HD_i} C_{tr_i}(\xi) \cos(\theta + \tau_i) d\xi - \int_{L_{w_i}} f_{HS_i} C_{tr_i}(\xi) d\xi;\ i = 1,\ 2 \tag{24}$$

## 2.5. Frictional Forces

The friction drag force entered the single-stepped planing hull, calculated with consideration of two terms. The first term acts on the pressure area and the second term acts on the spray area. The frictional drag force on the pressure area of the single-stepped planing hull can be calculated by measuring the wetted surface of each section for each planing surface as follows,

$$S_{P_i} = \int_{L_{w_i}}^{L_{c_i}} \frac{2c_i}{\cos(\beta_i + \beta_{L_i})} d\xi + \int_{L_{c_i}}^{\text{Transom or Step}} \frac{2B}{\cos(\beta_i + \beta_{L_i})} d\xi; \; i = 1, 2 \tag{25}$$

The frictional drag on the pressure area can be calculated using the following equation,

$$Df = 0.5\rho v^2 (Cf_1 S_{P_1} + Cf_2 S_{P_2}) \tag{26}$$

where $Cf_i$ is the frictional drag coefficient and calculated based on ITTC,1957 [26] as follows:

$$Cf_i = \frac{0.075}{(\log_{10}^{R_{n_i}} - 2)^2}; \; (i = 1, 2) \tag{27}$$

$$Rn_i = \frac{V\lambda_i}{v}; \; (i = 1, 2) \tag{28}$$

$$\lambda_i = \frac{Lc_i + Lw_i}{2}; \; (i = 1, 2). \tag{29}$$

The frictional drag on the spray area, $R_{s_i}$, has been calculated for each planing surface by the following equation,

$$R_{spray_i} = fs_i |\cos(2\alpha_i)|; \; i = 1, 2 \tag{30}$$

where $fs_i$ and $\alpha_i$ are calculated separately for each planing surface, as follows,

$$fs_i = \frac{\rho v^2 B^2 Cf_i}{8 \sin(2\alpha_i) \cos(\beta_i + \beta_{L_i})}; \; i = 1, 2 \tag{31}$$

$$\alpha_i = \tan^{-1}\left(\frac{(Lw_i - Lc_i)}{B}\right); \; i = 1, 2 \tag{32}$$

The frictional drag force can be calculated by,

$$D = \sum_{i=1}^{2} R_{spray_i} + Df_i \tag{33}$$

Components of this frictional drag force in each of the directions for each planing surface are formulated as,

$$D_{x_i} = -R_{spray_i} \cos(\theta + \tau_i) - Df_i \cos(\theta + \tau_i); \; i = 1, 2 \tag{34}$$

$$D_{z_i} = -R_{spray_i} \sin(\theta + \tau_i) - Df_i \sin(\theta + \tau_i); \; i = 1, 2 \tag{35}$$

## 2.6. Resistance and Thrust

Te resistance of the single-stepped planing boat is computed by the following equation,

$$R = \sum_{i=1}^{2} D_{x_i} + F_{x_i}; \; i = 1, 2 \tag{36}$$

In the end, the thrust is calculated. The required thrust force for the single-stepped planing hull can be calculated by the equation below.

$$T = \frac{-\sum_{i=1}^{2} D_{x_i} + F_{x_i}}{\cos(\theta + \theta_T)}. \tag{37}$$

## 2.7. Computational Procedure

After presenting the mathematical model and its formulation, it is necessary to develop a computational procedure for solving Equations (1) and (2). To solve the equilibrium equations, a nonlinear optimization algorithm has been utilized, as shown in Figure 6. A computational procedure has been established for the prediction of the performance of the single-stepped planing boat as shown in Figure 6. The solution procedure for optimization is based on the constrained minimization of Equations (1) and (2) as an objective function. For this purpose, the Matlab command fmincont is applied in the mathematical model to minimize the equilibrium equations for each planing surface of the single-stepped hull in which the limits of guessed values (i.e., the trim and wetted length) are considered as inputs. In the end, by using the fmincon command, the trim and wetted length of the single-stepped hull can be obtained and both the heave and pitch equations would be solved.

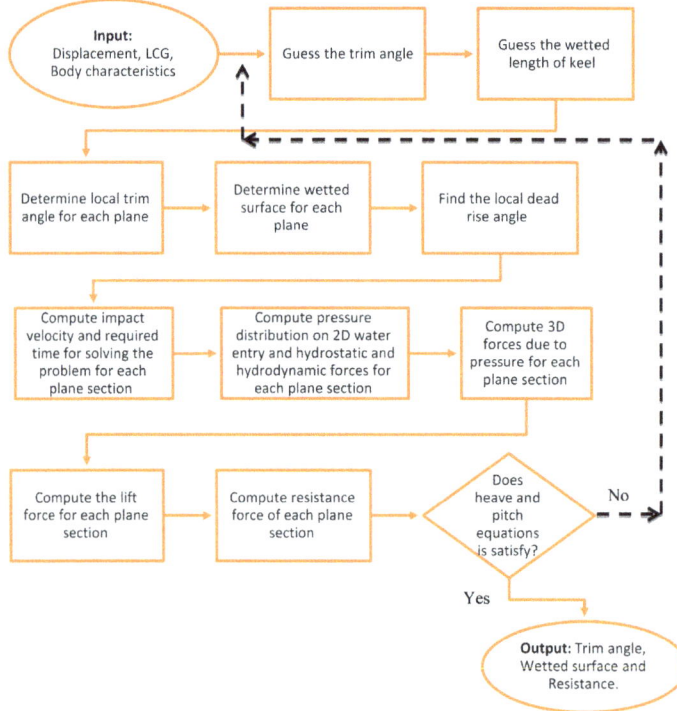

**Figure 6.** The computational workflow for determining the pressure distribution of the single-stepped planing hull.

## 3. Validation and Results

### 3.1. Model Tested and Experimental Details

The model used in this study represents an example of a modern high-speed hull for Rigid Inflatable Boats (RIB). This hull can be a representative hull for typical pleasure or military high-speed crafts. This model is hull number C03 and is one of the eight models of an unpublished systematic series. The body plan of the C03 hull is available in Figure 7.

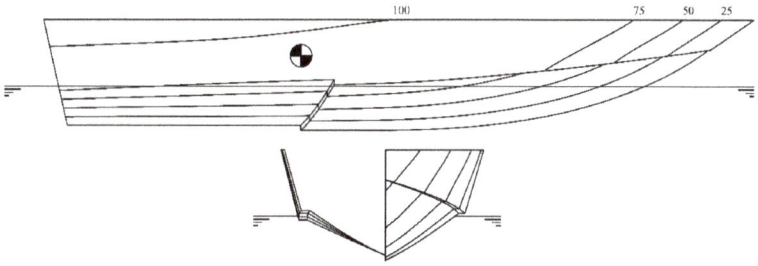

**Figure 7.** The C03 model body plan (transversal section every 0.100 m) and profile (buttock line every 0.025 m).

The parent hull for this research is a RIB built by MV Marine S.r.l. (Shipyard in Nola, Italy), type Mito 31, powered by outboard engines. The model is a hard chine hull with one transverse step, located in the same longitudinal position of the center of gravity with a forward-V shape (Figure 7). The model scale has the same main dimensions (keel line, chine line, deadrise angle, displacement, Longitudinal Center of Gravity (LCG), step shape, step angle) of RIB Mito 31, at a 1:10 scale. Details of the hull model scale are reported in Table 1.

**Table 1.** The C03 model details.

| Description | |
|---|---|
| Length overall: $L_{OA}$ (m) | 0.935 |
| Breadth max: $B_{MAX}$ (m) | 0.335 |
| Deadrise angle at transom (°) | 23 |
| Step height (mm) | 6 |
| Displacement (N) | 30.705 |
| LCG (%L) | 33 |
| Model scale | 1:10 |

The physical model for the towing tank tests was manufactured in hand-made layup through a mold, which was designed in 3D CAD/CAM, was milled with a CNC 5-Axes machine, and was built in FRP in accordance to ITTC [26], in fact, the model hull tolerances for breadth, drought, and length are ±0.5 mm. The manufacturing tolerance for length is less than 0.05%, and special attention was paid into the shaping of chines, keel, transom, and step. The model was built with composite materials with a transparent bottom built only with isophthalic resin to provide a full view of the water flow under the hull.

The tests were performed in the towing tank at the marine engineering section of the Department of Industrial Engineering (DII) of the Università degli Studi di Napoli "Federico II". The main dimensions of the towing tank are length 136.0 m, width 9.0 m, and depth 4.5 m. Calm water resistance experiments were conducted with the down-thrust (DT) methodology proposed in Vitiello and Miranda [27] at the following Froude numbers ($Fr$): 0.866, 1.151, 1.702, 1.973, 2.330, 2.683, and 2.958. In the case of a model being small and light, the DT measurement solution is due to the high sensitivity of the hull model to the externally applied forces, i.e., the instrumentation weight.

The DT solution releases the tested model from the instrumentation weight, which, in many cases, is similar to the model weight and promotes higher accuracy in measurements of resistance, sinkage, and trim. This experimental method has proven to reproduce the real system of forces exerted by outboard engines.

In fact, the engines, when going forward, transfer T thrust to the transom through forces applied in the lowest bracket area. Consequently, the system forces of the two outboard engines and hull are similar to two beams supported by a pin and a roller. The DT resistance methodology considers that, in a horizontal position, and in a trim angle at rest equal at zero, the direction of the thrust force is applied in point P, i.e., the intersection between the projection of the engine thrust direction on a keel plane and keel line at the bow, as shown in Figure 8.

**Figure 8.** The down-thrust method: towing tank thrust and true force system applied on the hull.

### 3.2. Towing Tank vs. 2D + T Method Results

The C03 towing tank tests were used to validate the 2D + T analytical model developed; in this paragraph, a comparison analysis between the experimental and analytical method is done. In Table 2 the values of non-dimensional resistance, the dynamic trim angle, and non-dimensional wetted surface are shown.

**Table 2.** The comparison between the experimental and analytical results.

| Fr | $RT_M/\Delta$ | | | Trim | | | $WS/\nabla^{2/3}$ | | |
|---|---|---|---|---|---|---|---|---|---|
| | Exp. | 2D + T Approach | Error | Exp. | 2D + T Approach | Error | Exp. | 2D + T Approach | Error |
| | (-) | (-) | (%) | (deg) | (deg) | (%) | (-) | (-) | (%) |
| 0.866 | 0.182 | 0.181 | 0.2 | 3.550 | 4.500 | −26.8 | 6.63 | 3.43 | 48.3 |
| 1.151 | 0.208 | 0.201 | 3.1 | 4.420 | 3.755 | 15.0 | 4.85 | 2.99 | 38.4 |
| 1.702 | 0.261 | 0.255 | 2.3 | 3.270 | 2.880 | 11.9 | 3.88 | 2.47 | 36.2 |
| 1.973 | 0.318 | 0.288 | 9.5 | 2.870 | 2.605 | 9.2 | 3.54 | 2.31 | 34.7 |
| 2.330 | 0.415 | 0.336 | 19.1 | 2.690 | 2.326 | 13.5 | 3.32 | 2.15 | 35.2 |
| 2.683 | 0.501 | 0.389 | 22.3 | 2.520 | 2.113 | 16.2 | 3.23 | 2.03 | 37.3 |
| 2.958 | 0.566 | 0.434 | 23.5 | 2.580 | 1.976 | 23.4 | 2.85 | 1.94 | 31.7 |

The uncertainty bars in Figures 9–11 are in accordance with the experimental uncertainty evaluation reported in De Marco et al. [3].

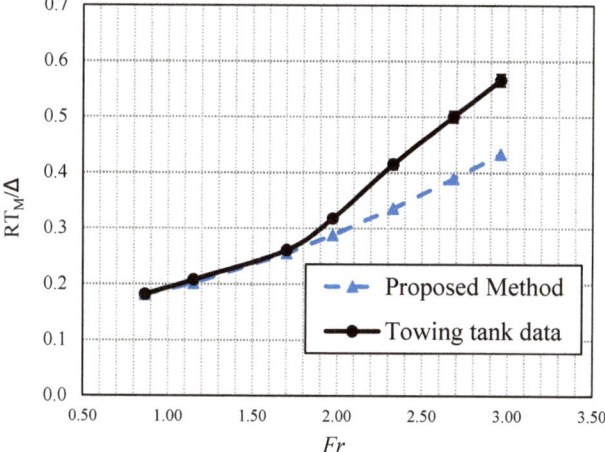

**Figure 9.** The non-dimensional resistance: 2D + T method vs. experimental results.

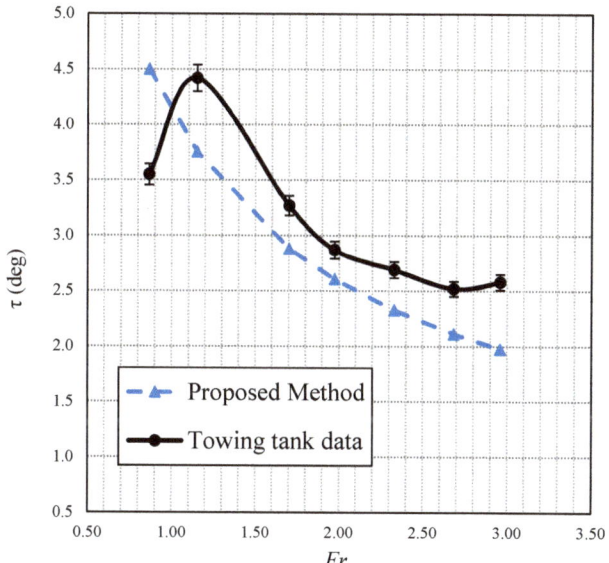

**Figure 10.** The dynamic trim angle: 2D + T method vs. experimental results.

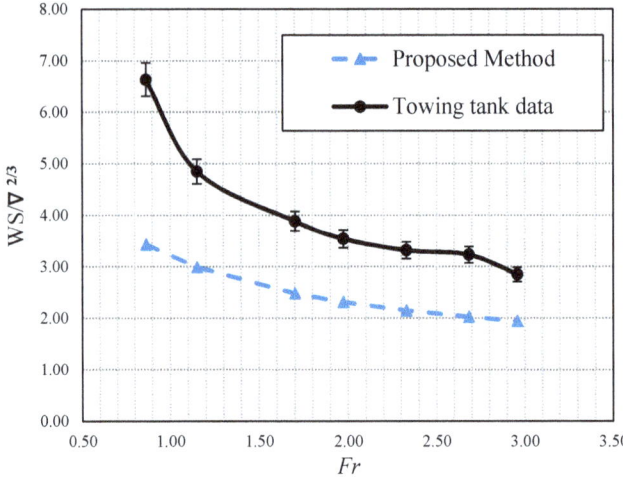

**Figure 11.** The dynamic wetted surface: 2D + T method vs. experimental results.

Observing the results, it is possible to observe that the non-dimensional resistance comparison error increases by increasing the $Fr$—but until $Fr = 1.973$ is less the 10% and the Mean Squared Error (MSE) is equal to 13.7%.

For the dynamic trim angle, the comparison error shows a minimum for the intermediate values of $Fr$ (1.702, 1.973, and 2.330), and the error increases for extreme values, where the trim angles are, respectively, the minimum and maximum. The MSE for the dynamic trim angle is equal to 16.4%.

For the non-dimensional wetted surface, the trend of the error decreases, thus increasing the $Fr$, but the value is constantly greater than the 30%, and the MSE is equal to 35.3%.

Furthermore, in order to evaluate the effectiveness of the 2D + T approach, the 2D + T results are compared with the Computational Fluid Dynamic (CFD) analysis performed for the same hull model available, with all CFD simulation details, in De Marco et al. [3]. Table 3 shows that URANS high-quality simulations can achieve results close to the towing tank tests with an error generally less than 10% for the dynamic trim angle and total resistance. However, for the wetted surface, the absolute values of error are comparable, particularly for the highest $Fr$.

**Table 3.** CFD vs. 2D + T results: comparison errors.

| | $RT_M/\Delta$ | | Trim | | $WS/\nabla^{2/3}$ | |
|---|---|---|---|---|---|---|
| $Fr$ | EXP–CFD | EXP–2D + T | EXP–CFD | EXP–2D + T | EXP–CFD | EXP–2D + T |
| | (%) | (%) | (%) | (%) | (%) | (%) |
| 0.866 | 5.46 | 0.20 | −9.01 | −26.76 | −8.14 | 48.33 |
| 1.702 | −1.90 | 2.32 | −1.22 | 11.93 | −31.20 | 36.24 |
| 2.330 | 9.33 | 19.07 | −0.37 | 13.53 | −31.15 | 35.19 |
| 2.958 | 5.26 | 23.46 | −3.10 | 23.41 | −36.13 | 31.71 |

### 3.3. Wetted Surfaces and Wetted Length Analysis

The experimental wetted surface values are estimated through the digital analysis of video frames, which are referenced to the original 3D CAD model, as shown in Figure 12. The analytical values are calculated according to the computational workflow, as shown in Figure 6. As previously mentioned (Figure 11), the comparison error between the experimental and analytical results substantially decreases, thus increasing the $Fr$. However, the comparison error is considerable in the whole $Fr$ range.

Hence, a deeper analysis is required in order to investigate the issues in wetted surface evaluation by comparing the two different wetted surface evaluations, as shown in Table 4.

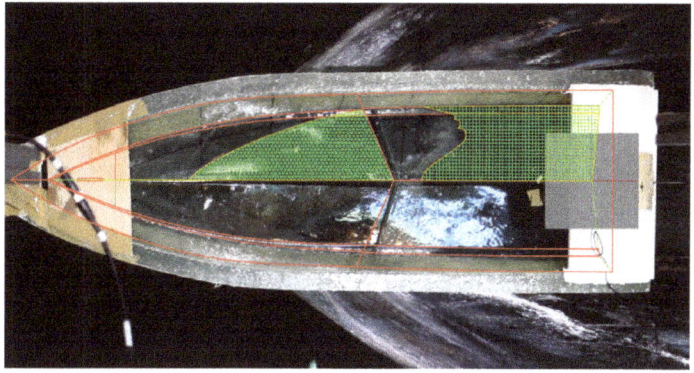

**Figure 12.** The experimental wetted surface evaluation using recorded video frame @$Fr$ = 1.151.

As opposed to what happens in the URANS simulations that, generally, overestimated the wetted surface, as stated in De Marco et al. [3], De Luca et al. [28], and Mancini et al. [29], the 2D + T method underestimates the wetted surface values. In particular, at low $Fr$, the analytical method fails in the estimation, specifically in the aft wetted surface evaluation. By increasing the $Fr$, the air cavity behind the step increases, and the wetted surface became narrower; thus the comparison error value is reduced. However, the 2D + T approach is not able to predict the unsteady turbulent phenomena that characterize the hydrodynamic flow behind the step. Another shortcoming of the 2D + T method is the lack of capability of the developed analytical approach to consider the transversal step angle.

These issues also affect the dynamic wetted length evaluation. Thus, the dynamic wetted length computed by the 2D + T method is significantly less than the experimental one as it is possible to observe in Figure 13. The dynamic wetted length is strictly linked to the air cavity generated behind the step. The difference in the air cavity evaluation can be detected in Table 4, in particular for low $Fr$ values.

**Table 4.** The detailed view of the dynamic wetted surface for the experimental test and analytical method.

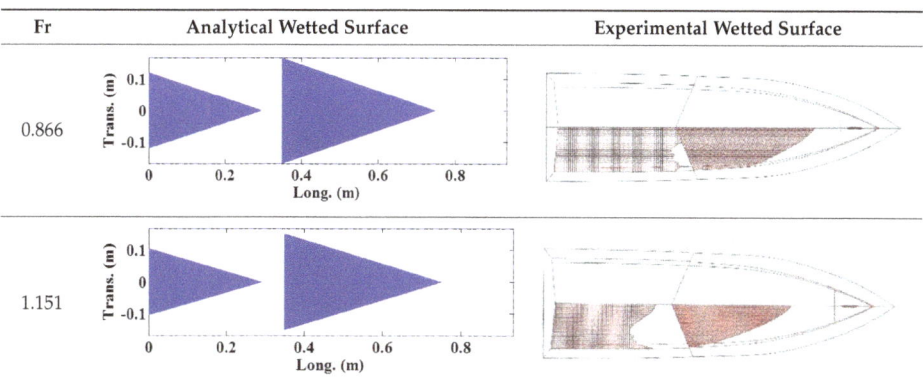

Table 4. *Cont.*

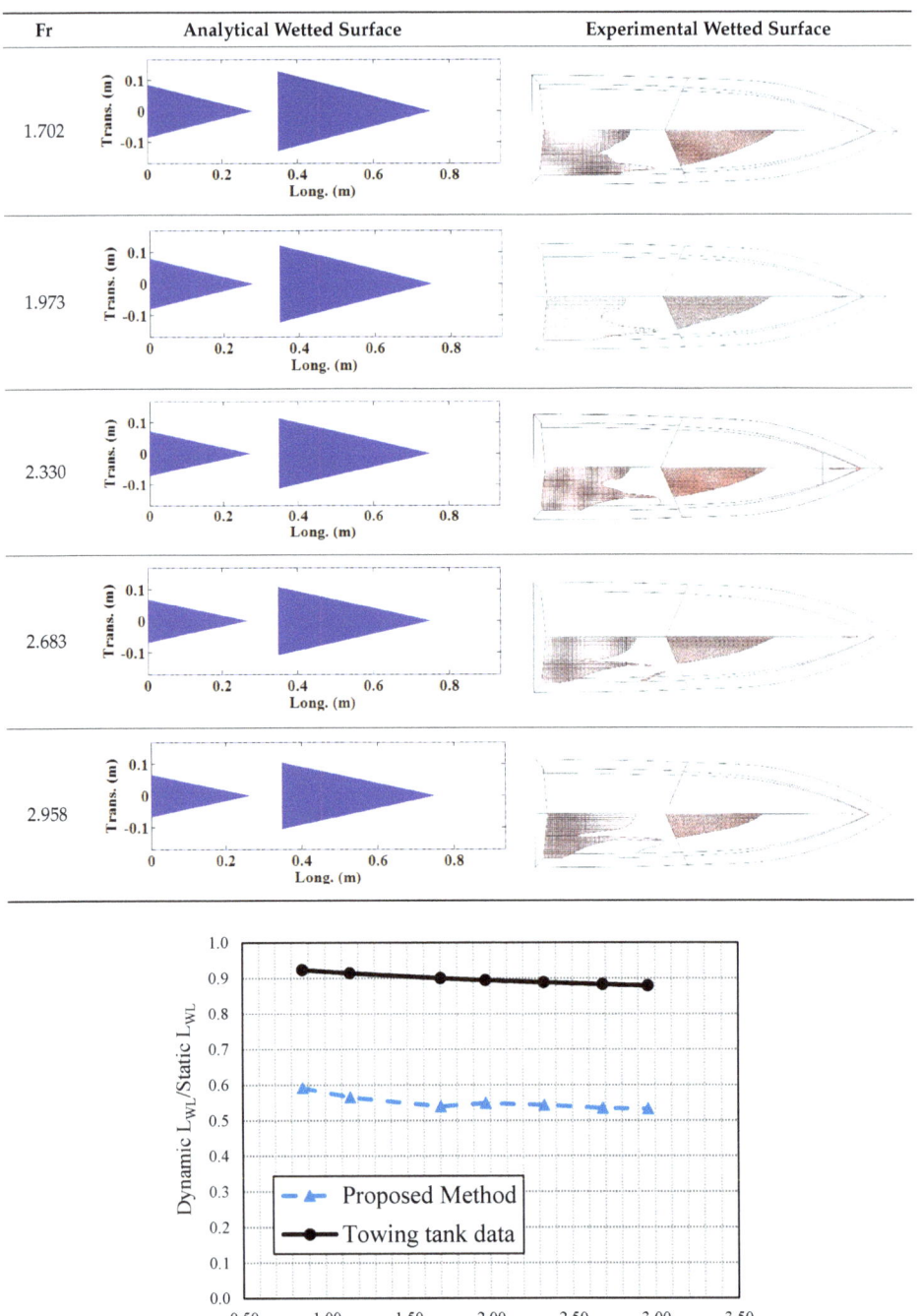

Figure 13. The dynamic wetted surface: 2D + T method vs. experimental results.

## 4. Conclusions

A 2D + T analytical method has been developed for the performance evaluation of stepped hulls. In order to validate the mathematical model, calm water resistance experiments in a towing tank on a single stepped hull model with a transparent bottom have been used.

The comparison between the analytical approach and towing tank test results shows an acceptable reliability for the resistance and dynamic trim evaluation, in particular for a Froude number up to 2.0, with a Mean Squared Error equal to 13.7% for resistance and 16.4% for dynamic trim. However, the dynamic wetted surface, as well as the dynamic wetted length evaluation, presents a larger error, in particular at low Froude numbers. This issue can be related to the two main shortcomings of the 2D + T approach, as the inability to describe the unsteady turbulent phenomena behind the step and the inability to take into account the transversal step angle. The last issue can be overcome by changing the mathematical code. However, the unsteady turbulent phenomena cannot be predicted by the 2D + T approach. These phenomena can be predicted by experimental or numerical means, considering the URANS/LES simulations. The accuracy of the CFD method and towing tank test is high but there are many complexities due to the simulation setup, the experimental arrangement, the high computational effort required, and the high cost. Hence, these performance evaluation methods cannot easily and quickly be used in particular in the early design stage. Therefore, the 2D + T method is more cost-effective for the designers at the first design stage in order to quickly assess a stepped hull shape, the power prediction, and the dynamic trim angle, thus defining the main hull parameters.

Promising results of the current study signals that the 2D + T theory also has a suitable accuracy in motion prediction of single-stepped planing hulls that can aid engineers in the early stage design process of a stepped planing hull. The method can be considered a very fast tool to provide the results in the concept design stage. Additionally, it can further develop in terms of the accuracy of the high Froude numbers and in order to implement the capability to give output in terms of seakeeping, maneuvering, and steady drift tests by considering other motions for the wedge. Therefore, future studies will focus on the further extension of this method for sweep-back step and by considering other motions.

**Author Contributions:** R.N.B. and S.M. conceived of the presented idea. R.N.B. developed the theory and performed the computations. L.V. and S.M. conceived, planned the experiments, and carried out the experiments. R.N.B. and S.M. verified the analytical methods. M.D.C., L.V., and S.M. supervised the findings of this work. R.N.B., S.M., and M.D.C. wrote the paper with input from all authors.

**Funding:** This work has been supported in part by ECO-RIB project grant (D.M. 01/06/2016–Horizon 2020–PON 2014/2020), and Vincenzo Nappo of the MV Marine S.r.l.

**Conflicts of Interest:** The authors declare no conflict of interest. The funders had no role in the design of the study; in the collection, analyses, or interpretation of data; in the writing of the manuscript, or in the decision to publish the results.

## Nomenclature

*Boat Characteristics*

| | |
|---|---|
| $B$ | Beam of the boat (m) |
| $Cf_i$ | Frictional coefficient of the $i$th body |
| $Fr$ | Froude number |
| $Fr_B$ | Beam Froude number |
| $H_{step_1}$ | Height of step (m) |
| $L$ | Length of the boat (m) |
| $Lc_i$ | Chine wetted length of the $i$th body (m) |
| $LCG$ | Longitudinal Center of Gravity (m) |
| $Lw_i$ | Wetted length of body $i$th body (m) |
| $Rn_i$ | Reynolds Number of the $i$th body |
| $S_{P_i}$ | Wetted area of the $i$th body |

*Boat Characteristics*

| | |
|---|---|
| $V$ | Forward moving velocity of the boat (m s$^{-1}$) |
| $\alpha_i$ | stagnation line angle of the $i$th body |
| $\beta_i$ | Deadrise angle of the boat |
| $\beta_{L_i}$ | Local deadrise angle of the boat of the $i$th body |
| $\Delta$ | Weight of boat (N) |
| $\lambda_i$ | Mean wetted length of the $i$th body |
| $\tau_i$ | Local trim angle of the $i$th body |
| $\theta$ | Dynamic trim angle of the hull |

*Distance*

| | |
|---|---|
| $a_{non}$ | Non-dimensional distance at which transom reduction appears |
| $L_s$ | Distance of step from the transom (m) |
| $L_{dry}$ | Dry length of step from the transom |
| $x, y, z$ | Longitudinal (positive forward), transverse (positive starboard), and vertical distances (positive downward) from CG (Oxyz) (m) |
| $\xi, \eta, \zeta$ | Longitudinal (positive forward), transverse (positive starboard), and vertical distances (positive downward) (m) |
| $\xi_i'$ | Distance of section from the step or transom just located behind the section (m) |
| $\zeta_{s_i}$ | Distance of section from intersection of the keel and calm water of the $i$th body (m) |

*Force and Moments*

| | |
|---|---|
| $Df$ | Frictional drag on pressure area (N) |
| $F_i$ | Pressure force on $i$th body (N) |
| $f_{s_i}$ | Drag acting on the spray area (N) |
| $R$ | Total resistance of the vessel |
| $R_{spray_i}$ | frictional drag of Whisker spray of the $i$th body |
| Subscript $x$ | Force component in surge direction (N) |
| Subscript $z$ | Force component in heave direction (N) |
| Subscript $\theta$ | Force component in pitch direction (N) |

*Physical Parameters*

| | |
|---|---|
| $g$ | Gravitational constant |
| $P_i$ | Pressure of the $i$th body (Pa) |
| $\rho$ | Fluid density (kg m$^{-3}$) |

*Sectional Parameters Related to 2.5D Theory*

| | |
|---|---|
| $A_i$ | Submerged area of the $i$th body (N m$^{-1}$) |
| $c_i$ | Half beam of spray in transverse plane (m) |
| $\dot{c}_i$ | Time derivation of c (m s$^{-2}$) |
| $C_{tr_i}$ | Transom reduction at the section of the $i$th body (N m$^{-1}$) |
| $f_{HD_i}$ | Hydrodynamic force of each section of the $i$th body (N m$^{-1}$) |
| $f_{B_i}$ | Hydrostatic force of each section of the $i$th body (N m$^{-1}$) |
| $l$ | Distance from wedge apex in the direction of wedge wall (m) |
| $t$ | Time |
| $t_{cw_i}$ | Chine wetting time of the $i$th body (s) |
| $t_{p_i}$ | Solution time for water entry problem of the $i$th body |
| $w_i$ | Impact velocity of the $i$th body |
| $y_i$ | Lateral distance from wedge apex of the $i$th body |
| Subscript $H$ | component in horizontal direction |
| Subscript $V$ | component in vertical direction |

## References

1. Taunton, D.J.; Hudson, D.A.; Shenoi, R.A. Characteristics of a Series of High-speed Hard Chine Planing Hulls Part 1: Performance in Calm Water. *Int. J. Small Craft Technol.* **2010**, *152*, B55–B75.
2. Lee, E.; Pavkov, M.; Mccue Weil, W. The Systematic Variation of Step Configuration and Displacement for a Double-Step Planing Craft. *J. Ship Prod. Des.* **2014**, *30*, 89–97. [CrossRef]

3. De Marco, A.; Mancini, S.; Miranda, S.; Vitiello, L.; Scognamiglio, R. Experimental and numerical hydrodynamic analysis of a stepped planing hull. *Appl. Ocean Res.* **2017**, *64*, 135–154. [CrossRef]
4. Savitsky, D. Hydrodynamic Design of Planing Hull. *Mar. Technol.* **1964**, *1*, 71–95.
5. Dashtimanesh, A.; Tavakoli, S.; Sahoo, P. A simplified method to calculate trim and resistance of a two-stepped planing hull. *Ships Offshore Struct.* **2017**, *12* (Suppl. 1), S317–S329. [CrossRef]
6. Niazmand Bilandi, R.; Dashtimanesh, A.; Tavakoli, S. Development of a 2D + T theory for performance prediction of double-stepped planing hulls in calm water. *J. Eng. Mar. Environ.* **2018**. [CrossRef]
7. Di Caterino, F.; Niazmand Bilandi, R.; Mancini, S.; Dashtimanesh, A.; De Carlini, M. Numerical Way for a Stepped Planing Hull Design and Optimization. In Proceedings of the 19th International Conference on Ship & Maritime Research, Trieste, Italy, 20–22 June 2018.
8. Dashtimanesh, A.; Esfandiari, A.; Mancini, S. Performance Prediction of Two-Stepped Planing Hulls Using Morphing Mesh Approach. *J. Ship Prod. Des.* **2018**, 1–13. [CrossRef]
9. Von Karman, T. *The Impact on Seaplane Floats during Landing (National Advisory Committee for Aeronautics No. 321)*; NACA Translation: Washington, DC, USA, 1929.
10. Wagner, H. *Phenomena Associated with Impacts and Sliding on Liquid Surfaces*; NACA Translation: Washington, DC, USA, 1932.
11. Svahn, D. Performance Prediction of Hulls with Transverse Steps. Master's Thesis, Marina System Centre for Naval Architecture, KTH University, Stockholm, Sweden, 2009.
12. Savitsky, D.; Morabito, M. Surface wave contours associated with the fore body wake of stepped planing hulls. *Mar. Technol.* **2010**, *47*, 1–16.
13. Danielsson, J.; Strømquist, J. Conceptual Design of a High Speed Superyacht Tender Hull Form Analysis and Structural Optimization. Master's Thesis, Marina System Centre for Naval Architecture, KTH University, Stockholm, Sweden, 2012.
14. Zarnick, E. *A Nonlinear Mathematical Model of Motions of a Planing Boat in Regular Waves Technical Report*; Report No. DTNSRDC-78/032; Bethesda, David W Taylor Naval Ship Research and Development Center: Rockville, MD, USA, 1978.
15. Ghadimi, P.; Dashtimanesh, A.; Djeddi, S.R.; Maghrebi, Y.F. Development of a mathematical model for simultaneous heave, pitch and roll motions of planing vessel in regular waves. *Int. J. Sci. World* **2013**, *1*, 44–56. [CrossRef]
16. Ghadimi, P.; Dashtimanesh, A.; Faghfoor Maghrebi, Y. Initiating a mathematical model for prediction of 6-DOF motion of planing crafts in regular waves. *Int. J. Eng. Math.* **2013**, *2013*, 853793. [CrossRef]
17. Ghadimi, P.; Tavakoli, S.; FeiziChekab, M.A.; Dashtimanesh, A. Introducing a Particular Mathematical Model for Predicting the Resistance and Performance of Prismatic Planing Hulls in Calm Water by Means of Total Pressure Distribution. *J. Nav. Arch. Mar. Eng.* **2015**, *12*, 73–94. [CrossRef]
18. Ghadimi, P.; Tavakoli, S.; Dashtimanesh, A. Coupled heave and pitch motions of planing hulls at non-zero heel angle. *Appl. Ocean Res.* **2016**, *59*, 286–303. [CrossRef]
19. Ghadimi, P.; Tavakoli, S.; Dashtimanesh, A. An analytical procedure for time domain simulation of roll motion of the warped planing hulls. *Proc. Inst. Mech. Eng. Part M J. Eng. Mar. Environ.* **2016**, *230*, 600–615. [CrossRef]
20. Tavakoli, S.; Ghadimi, P.; Dashtimanesh, A.; Sahoo, P. Determination of hydrodynamic coefficients in roll motion of high-speed planing hulls. In Proceedings of the 13th International Conference on Fast Sea Transportation, Washington, DC, USA, 1–4 September 2015.
21. Tavakoli, S.; Ghadimi, P.; Dashtimanesh, A. A nonlinear mathematical model for coupled heave, pitch, and roll motions of a high-speed planing hull. *J. Eng. Math.* **2017**, *104*, 157–194. [CrossRef]
22. Tavakoli, S.; Ghadimi, P.; Sahoo, P.K.; Dashtimanesh, A. A hybrid empirical–analytical model for predicting the roll motion of prismatic planing hulls. *Proc. Inst. Mech. Eng. Part M J. Eng. Mar. Environ.* **2018**, *232*, 155–175. [CrossRef]
23. Tavakoli, S.; Dashtimanesh, A.; Sahoo, P.K. An oblique 2D + T approach for hydrodynamic modeling of yawed planing boats in calm water. *J. Ship Prod. Des.* **2017**. [CrossRef]
24. Algarin, R.; Tascon, O. Hydrodynamic modeling of planing boats with asymmetry and steady condition. In Proceedings of the 9th International Conference on High Performance Marine Vehicles (HIPER 11), Naples, Italy, 25–27 May 2011.

25. Garme, K. Improved time domain simulation of planing hulls in waves by correction of the near-transom lift. *Int. Shipbuild. Prog.* **2005**, *52*, 201–230.
26. The International Towing Tank Conference. *ITTC Recommended Procedures and Guidelines: Ship Models*; ITTC: Rio de Janeiro, Brazil, 2011.
27. Vitiello, L.; Miranda, S. Propulsive performance analysis of a stepped hull by model test results and sea trial data. In Proceedings of the 10th Symposium on High Speed Marine Vehicles (HSMV 2014), Naples, Italy, 15–17 October 2014; ISBN 9788890611216.
28. De Luca, F.; Mancini, S.; Pensa, C.; Miranda, S. An Extended Verification and Validation Study of CFD Simulations for Planing Hulls. *J. Ship Res.* **2016**, *60*, 101–118. [CrossRef]
29. Mancini, S.; De Luca, F.; Ramolini, A. Towards CFD guidelines for planing hull simulations based on the Naples Systematic Series. In Proceedings of the Computational Methods in Marine Engineering VII (Marine 2017), Nantes, France, 15–17 May 2017.

© 2018 by the authors. Licensee MDPI, Basel, Switzerland. This article is an open access article distributed under the terms and conditions of the Creative Commons Attribution (CC BY) license (http://creativecommons.org/licenses/by/4.0/).

*Article*

# Investigating the Pre-Damaged PZT Sensors under Impact Traction

Sakineh Fotouhi [1,*], Mohamad Fotouhi [2], Ana Pavlovic [3] and Nenad Djordjevic [4]

1. Mechanical Engineering Department, University of Tabriz, Tabriz 51666-14766, Iran
2. Department of Design and Mathematics, University of the West of England, Bristol BS16 1QY, UK; mohammad.fotouhi@uwe.ac.uk
3. Department of Industrial Engineering, University of Bologna, Viale Risorgimento 2, 40136 Bologna, Italy; ana.pavlovic@unibo.it
4. Institute of Materials and Manufacturing, College of Engineering, Design and Physical Sciences, Brunel University London, UB8 3PN London, UK; nenad.djordjevic@brunel.ac.uk
* Correspondence: s.fotouhi@tabrizu.ac.ir; Tel.: +98-4474-0726-6470

Received: 2 October 2018; Accepted: 11 November 2018; Published: 19 November 2018

**Abstract:** Ships are usually under vibration, impact, and other kinds of static and dynamic loads. These loads arise from water flow across the hull or surfaces, the propeller cavitation, and so on. For optimal design purposes and reliable performance, experimental measurements are necessary. These sensors are often used under or near the water, working conditions that improve the risk of sensor damage. This paper aims at investigating, by the use of finite elements, the behavior of damaged piezoelectric sensors under traction and impact loads. The numerical method was calibrated using results available in the literature regarding piezoelectric and elastic plates with a central crack. After calibration, the simulation was used on two types of Lead-Zirconium-Titanium oxide (PZT) sandwich panel structures reinforced by aluminum skins. The results proved that the damage size and impact energy are important factors affecting the response of piezoelectric sensors; therefore, special attention might be considered when using these sensors for marine applications.

**Keywords:** piezoelectric sensor; damaged sensor; impact traction; Lead-Zirconium-Titanium (PZT); fracture mechanics; marine industry

---

## 1. Introduction

Lead-Zirconium-Titanium oxide (PZT) is a piezoelectric ceramic often used for actuator and sensor applications. PZTs have been widely used in underwater transducers and sensors. PZTs have sensory application in marine and other engineering application industries [1–6]. However, PZTs are very brittle and susceptible to damage during service [7]. Therefore, it is important to understand the fracture behavior of these materials with the existence of damage, and also in comparison with other conventional techniques [8–10]. As PZT ceramics are brittle and sensitive, they are usually made in sandwich panel (or composite) forms. There are many studies about the fracture behavior of the piezoelectric composites. In References [11–15], antiplane strain of damaged PZT strips and composites were investigated. Recently, there have been some applicable analytical solutions for the fracture behavior in PZTs. For example, Shindo et al. [16] analyzed an infinite orthotropic PZT plate with a Griffith crack under shear impact loading. Chen and Meguid [17] studied a cracked piezoelectric strip influenced by electromechanical loading. Wanga and Noda [18] studied a piezoelectric layer bonded to the surface of an elastic structure with a crack under transient load in a piezoelectric layer bonded to a dissimilar elastic layer under transient load. Ueda [19] studied the dynamic response of a central cracked piezoelectric composite plate with impact loading. Garcıa-Sanchez et al. [20,21] used a Boundary Element Method (BEM) for analyses of the transient response of cracked linear piezoelectric solids. Fotouhi et al. used Finite Element Modelling (FEM) to investigate the thickness and material properties of piezoelectric layer influence on the fracture response of PZT composite. The results proved that the stress intensity factor is influenced by these parameters [22].

## 2. Materials and Methods

### 2.1. Aim and Scope

To improve the situational awareness, obstacle avoidance of marine vehicles and energy harvesting, thin-film piezoelectric pressure sensors are used widely for underwater sensing. Most of these piezoelectric sensors/actuators are manufactured with a sandwich structure [23]. This study aims to investigate the dynamic behavior of a sandwich piezoelectric panel, piezoelectric layers between two Aluminum skins, with a central crack under impact traction and dynamic stress intensity factor (DSIF) near the tip of the crack. The research analyzes, in particular, the effect of the pre-existing crack on the response of a piezoelectric layered composite plate with finite dimensions. The results are useful to (a) develop an optimum sensor design; (b) consider the crack effects on the piezoelectric behavior, and (c) verify if PZT sensors are suitable or not for the use.

### 2.2. Validation

The results were obtained using numerical simulations. At first, the computational model, developed in accordance with previous researches [22,24–26], was checked using two samples from literature: a pure piezoelectric plate of $BaTiO_3$ [27], and a pure elastic sample [28,29] with a crack. A brief characteristic of these cases is illustrated in Figure 1. The numerical results were compared with the available information, and an acceptable matching was observed (Table 1). The error percentage (Error (%)) is calculated by 100*((Our simulation value—Previous study's value)/Previous study's value)). Figure 2 illustrates the efficiency of the simulation method.

After the validation phase, the procedure was applied to simulate the finite PZT composite plates illustrated in Figure 3. Numerical results for the DSIF were observed for several piezoelectric layered composite plates. The main objective of this paper is to investigate the effect of impact energy level and a pre-existing crack length on fracture response of a PZT composite plate.

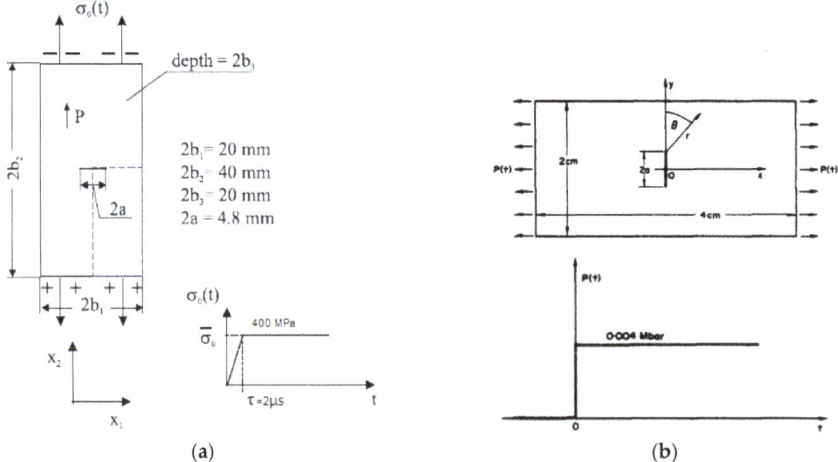

**Figure 1.** Schematic of the (a) piezoelectric ($BaTiO_3$) plate [26] and (b) steel plate [27].

**Table 1.** Comparison of the results (present results and previous works [22,24–26]) for the investigated case studies.

| BaTiO$_3$ Plate' Results Comprehension | K$_{max}$ (MPa) | T$_{max}$ (μs) | Steel Plate' Results Comprehension | K$_{max}$ (MPa) | T$_{max}$ (μs) |
|---|---|---|---|---|---|
| Obtained results | 1.15 | 9.5 | Obtained results | 2.559 | 7 |
| Enderlein's work | 1.15 | 9.5 | Chen's work | 2.698 | 6 |
| Error (%) | 0.41 | 0.115 | Error (%) | 5.13 | −6.06 |

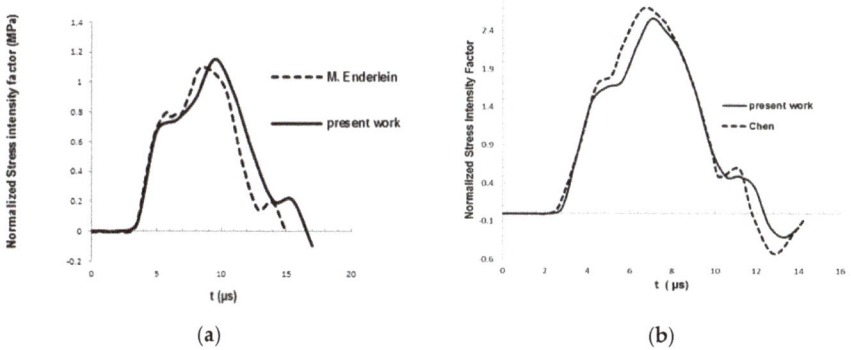

**Figure 2.** Dynamic stress intensity factor (DSIF) (k$_I$) versus time of (a) BaTiO$_3$ and (b) steel plates.

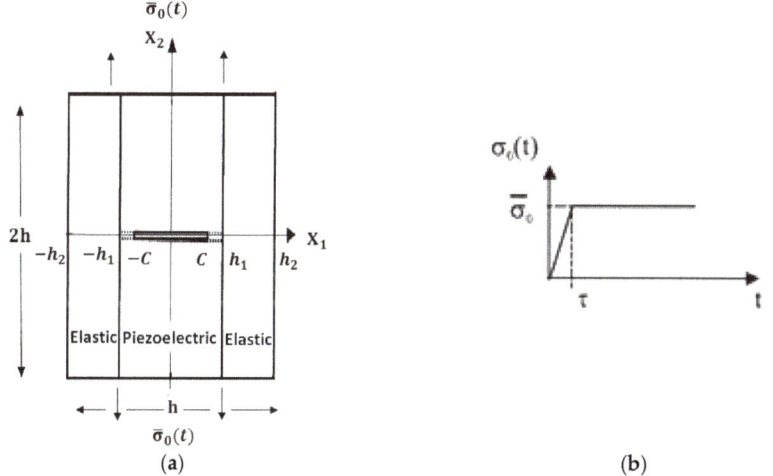

**Figure 3.** Schematic of (a) the Lead-Zirconium-Titanium oxide (PZT) cracked composite plate and (b) the applied stress.

### 2.3. Simulation of the Piezoelectric Composite Plate

In this article, PZT composite contains a PZT core layer of $2h_1$ thickness that interleaved between two elastic layers of thickness $h_2$–$h_1$ with a crack in the center as shown in Figure 3. These piezoelectric materials have a wide application in real marine structure as sensors/actuators. The crack length is 2c [19]. PZT composite is subjected to an impact traction of the form $\bar{\sigma}_0(t)$ (according to Figure 3), where $\tau$ is 2 μs. The width of the PZT composite is defined by h = 20 mm [29]. The $h_1$ is equal to 5 mm and the $\bar{\sigma}_0(t)$ and c changed in the simulation according to the aims. The needed materials' properties are in Table 2. The Poisson's ratio of aluminum is $\nu = 0.28$. In addition, all of these needed parameters are listed in Table 3.

Table 2. Materials' properties.

| Material | Elastic Stiffnesses ($\times 10^{10}$ N/m$^2$) | | | | Piezoelectric Coefficients (C/m$^2$) | | | Dielectric Constants ($\times 10^{-10}$ C/Vm) | |
|---|---|---|---|---|---|---|---|---|---|
| | $c_{11}$ | $c_{33}$ | $c_{44}$ | $c_{13}$ | $e_{31}$ | $e_{33}$ | $e_{15}$ | $\varepsilon_{11}$ | $\varepsilon_{33}$ |
| PZT-4 | 13.9 | 11.3 | 2.56 | 7.43 | −6.98 | 13.8 | 12.7 | 60 | 54.7 |
| PZT-5H | 12.6 | 11.7 | 2.3 | 8.41 | −6.5 | 23.3 | 17 | 150.4 | 130 |
| Al | 8.84 | 8.84 | 2.7 | 3.43 | 0 | 0 | 0 | - | - |
| BaTiO$_3$ | 15 | 14.6 | 4.4 | 6.6 | −4.35 | 17.5 | 11.4 | 98.7 | 112 |
| Steel (Elastic) | Density ($\rho$): 5000 kg/m$^3$ | | | | Poisson ratio ($\nu$): 0.3 | | | Young's modulus ($G$): 76.923 GPa | |

Table 3. Value of the piezoelectric plates' simulation parameters.

| $h_2$ (mm) | $h_1$ (mm) | $h$ (mm) | $c$ (mm) | $\bar{\sigma}_0(t)$ MPa | $\tau$ (µs) |
|---|---|---|---|---|---|
| 10 | 5 | 20 | variable | variable | 2 |

The size and property of the piezoelectric materials depend on their application. This case study assumes a damaged piezoelectric sensor in marine, ocean, and other underwater structures. To examine the effect of damaged sensors, the FEM is used. Two samples are considered: Aluminum/PZT-4/Aluminum and Aluminum/PZT-5H/Aluminum. The piezoelectric plate was simulated with a shell planer part as an anisotropic plate with properties listed in Table 2. A uniform traction load with a defined amplitude (according to Figure 3) was applied on the piezoelectric laminate. Free edge boundary condition was used for the plate's boundaries to simulate the real-world application of the sensors. Standard, linear, and plane strain mesh type were used for the modeling. Sweep quad-dominated mesh was utilized for the crack tips, and free quad-dominated mesh with medium mesh size was used for the other regions. The influence of two variations, i.e., the crack length and impact energy levels, on DSIF under impact loadings in the PZT composite plates, are reported in Table 3.

## 3. Results

### 3.1. The Crack Length of the Piezoelectric Composite Plate

DSIF and stress distribution for the PZT composites (Aluminum/PZT-4/Aluminum and Aluminum/PZT-5H/Aluminum) with $\frac{h_2}{h_1} = 2$ and $h_1 = 5$ mm and different crack lengths were analyzed. The obtained results are indicated in Figures 4–7.

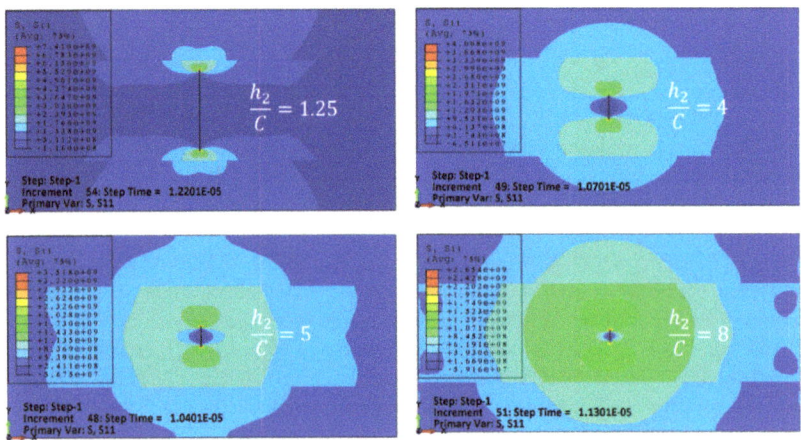

Figure 4. Crack length effect on stress distribution for PZT-4 composite.

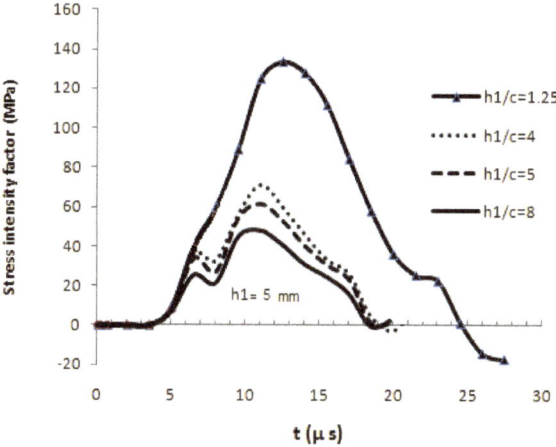

**Figure 5.** DSIF curves versus time of PZT-4 composite.

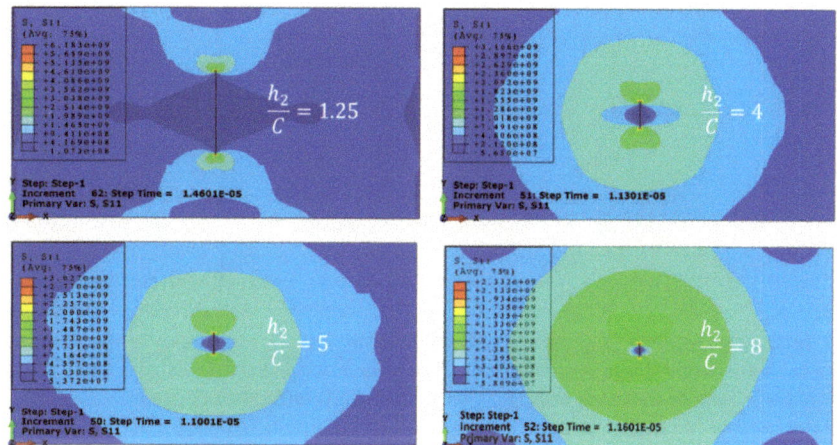

**Figure 6.** Crack length effect on stress distribution for PZT-5H composite.

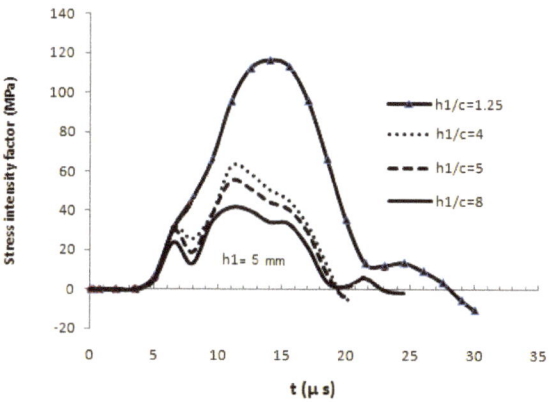

**Figure 7.** DSIF curves versus time of PZT-5H composite.

## 3.2. The Impact Energy Effect on the Piezoelectric Composite Plate

Piezoelectric sensors are used for marine applications for detection of the applied loads, stresses etc. These sensors usually are in different loads, such as impact, with different energy levels. To investigate the effect of impact energy level on stress distribution and DSIF, the piezoelectric composites with constant ratios of piezoelectric layer thickness ($h_1$) to crack length ($c$) and the composite thickness ($h_2$) to the piezoelectric layer thickness ($h_1$) equal 2 (i.e., $h_2/h_1 = 2$), were subjected to different loading conditions (100, 200, 300, and 400 MPa) and the results are shown in Figure 8. Some of the results show negative DSIF. It observed that when the crack is oriented vertically or almost vertically, the DSIF becomes negative, indicating that the crack closes because of compressive loading normal to the crack surface [30] and it is a good topic to investigate widely in the related study.

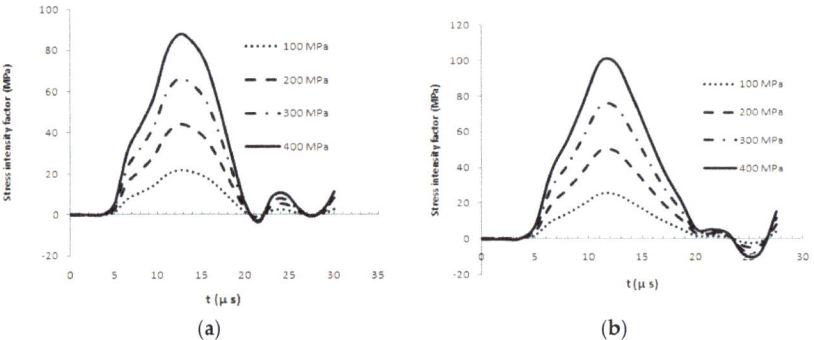

**Figure 8.** Effect of impact energy level on DSIF of (**a**) PZT-5H composite and (**b**) PZT-4 composite.

## 4. Discussion

The DSIF and $T_{max}$ increase by increasing the crack length. There are some fluctuations for higher numbers of $h_1/c$, but in reality, these situations are rarely observable and are not real case studies. It is also observed that by increasing the impact level the DSIF increases significantly, but there is almost no change in the $T_{max}$. This is in accordance with a usual fracture response of a laminated composite plate under impact or indentation load, where there is no difference between the rise time of the maximum load under different impact energy levels [19]. So, there is a good agreement between the fracture response of the investigated piezoelectric layered composites in this paper and conventional laminated composite plates with different stacking sequences, despite the multi-functionality of the piezoelectric composites.

## 5. Conclusions

This study aimed to investigate the effect of an existed damage in laminated PZT sensors that are often used under or near the water working conditions, i.e., the marine industry. For this reason, the fracture response of a PZT composite sensor with a central crack was considered under impact traction, which is a typical loading condition in the marine industry. The fracture response of the PZT composite sensor with a central crack, under impact traction, was considered. DSIF of the crack tip was analyzed. FEM was used to simulate the problem. The numerical results illustrated that DSIF and stress distribution of the PZT composite depends on the crack length and impact energy. The rise time of the stress intensity factor and stress distribution pattern changes with crack length, but the level of impact energy does not affect the rise time ($T_{max}$). It can be concluded that the proposed modeling procedure was very useful for the understanding of the dynamic behavior of piezoelectric materials with existent damage and can help to increase our knowledge for a better design and manufacturing of piezoelectric materials for engineering applications and more specifically marine industry.

**Author Contributions:** Conceptualization, S.F. and M.F.; methodology, M.F.; software, S.F. and A.P.; validation, M.F.; formal analysis, N.D.; investigation, S.F. and M.F.; resources, X.X.; data curation, A.P. and N.D.; writing-original draft preparation, S.F. and M.F.; writing-review and editing, S.F. and A.P.; supervision, N.D. and M.F.

**Funding:** This research received no external funding.

**Acknowledgments:** Special thanks to the Faculty of Environment and Technology, University of the West of England, for funding the researchers' mobility at the University of Bologna.

**Conflicts of Interest:** The authors declared no potential conflict of interest with respect to the research, authorship, and/or publication of this article.

## References

1. Huidong, L.; Zhiqun, D.D.; Yong, Y.; Thomas, J.C. Design Parameters of a Miniaturized Piezoelectric Underwater Acoustic Transmitter. *Sensors* **2012**, *12*, 9098–9109. [CrossRef]
2. Akdogan, E.K.; Allahverdi, M.; Safari, A. Piezoelectric composites for sensor and actuator applications. *IEEE Trans. Ultrason. Ferroelectr. Freq. Control* **2005**, *52*, 746–775. [CrossRef] [PubMed]
3. Galassi, C.; Roncari, E.; Capiani, C.; Fabbri, G.; Piancastelli, A.; Peselli, M.; Silvano, F. Processing of Porous PZT Materials for Underwater Acoustics. *Ferroelectrics* **2002**, *268*, 42–57. [CrossRef]
4. Fotouhi, M.; Saeedifar, M.; Yousefi, J.; Fotouhi, S. The application of an acoustic emission technique in the delamination of laminated composites. In *Focus on Acoustic Emission Research*; NOVA Publishers: Hauppauge, NY, USA, 2016.
5. Fotouhi, M.; Najafabadi, M.A. Acoustic emission-based study to characterize the initiation of delamination in composite materials. *J. Thermoplast. Compos. Mater.* **2016**, *29*, 519–537. [CrossRef]
6. Fotouhi, M.; Suwarta, P.; Jalalvand, M.; Czel, G.; Wisnom, M.R. Detection of fibre fracture and ply fragmentation in thin-ply UD carbon/glass hybrid laminates using acoustic emission. *Compos. Part A Appl. Sci. Manuf.* **2016**, *86*, 66–76. [CrossRef]
7. Mamoru, M.; Yuta, E.; Mitsuhiro, O. Fatigue life of piezoelectric ceramics and evaluation of internal damage. *Procedia Eng.* **2010**, *2*, 291–297.
8. Fragassa, C.; Minak, G. Measuring deformations in a rigid-hulled inflatable boat. *Key Eng. Mater.* **2017**, *754*, 295–298. [CrossRef]
9. Heidary, H.; Sadri, M.; Karimi, N.Z.; Fragassa, C. Numerical Study of Plasticity Effects in Uniform Residual Stresses Measurement by Ring-Core Technique. *J. Serb. Soc. Comput. Mech.* **2017**, *11*, 17–26. [CrossRef]
10. Zivkovic, I.; Pavlovic, A.; Fragassa, C.; Brugo, T. Influence of moisture absorption on the impact properties of flax, basalt and hybrid flax/basalt fiber reinforced green composites. *Compos. Part B Eng.* **2017**, *111*, 148–164. [CrossRef]
11. Shindo, Y.; Narita, F.; Tanaka, K. Electroelastic intensification near anti-plane shear crack in orthotropic piezoelectric ceramic strip. *Theor. Appl. Fract. Mech.* **1996**, *25*, 65–71. [CrossRef]
12. Shindo, Y.; Tanaka, K.; Narita, F. Singular Stress and electric fields of a piezoelectric ceramic strip with a finite crack under longitudinal shear. *Acta Mech.* **1997**, *120*, 31–45. [CrossRef]
13. Shindo, Y.; Watanabe, K.; Narita, F. Electroelastic analysis of a piezoelectric ceramic strip with a central crack. *Int. J. Eng. Sci.* **2000**, *38*, 1–19. [CrossRef]
14. Narita, F.; Shindo, Y. Layered piezoelectric medium with interface crack under anti-plane shear. *Theor. Appl. Fract. Mech.* **1998**, *30*, 119–126. [CrossRef]
15. Narita, F.; Shindo, Y. The interface crack problem for bonded piezoelectric and orthotropic layers under antiplane shear loading. *Int. J. Fract.* **1999**, *98*, 87–102. [CrossRef]
16. Shindo, Y.; Narita, F.; Ozawa, E. Impact response of a finite crack in an orthotropic piezoelectric ceramic. *Acta Mech.* **1999**, *137*, 99–107. [CrossRef]
17. Chen, Z.T.; Meguid, S.A. The transient response of a piezoelectric strip with a vertical crack under electromechanical impact load. *Int. J. Solids Struct.* **2000**, *37*, 6051–6062. [CrossRef]
18. Wang, B.L.; Noda, N.A. Crack in a piezoelectric layer bonded to a dissimilar elastic layer under transient load. *Arch. Appl. Mech.* **2001**, *71*, 487–494. [CrossRef]
19. Ueda, S. Impact response of a piezoelectric layered composite plate with a crack. *Theor. Appl. Fract. Mech.* **2002**, *38*, 221–242. [CrossRef]

20. Garcıa-Sanchez, F.; Zhang, Ch.; Sladek, J.; Sladek, V. 2D transient dynamic crack analysis in piezoelectric solids by BEM. *Comput. Mater. Sci.* **2007**, *39*, 179–186. [CrossRef]
21. Garcıa-Sanchez, F.; Zhang, C.; Saez, A. 2-D transient dynamic analysis of cracked piezoelectric solids by a time-domain BEM. *Comput. Methods Appl. Mech. Eng.* **2008**, *197*, 3108–3121. [CrossRef]
22. Fotouhi, S.; Khalili, M.R.S. Analysis of dynamic stress intensity factor of finite piezoelectric composite plate under a dynamic load. *FME Trans.* **2016**, *44*, 348–352. [CrossRef]
23. Safari, A. Development of Piezoelectric Composites for Transducers. *J. Phys. III* **1994**, *4*, 1129–1149. [CrossRef]
24. Fragassa, C.; De Camargo, F.V.; Pavlovic, A.; Silveira, A.C.F.; Minak, G.; Bergmann, C.P. Mechanical Characterization of Grés Porcelain and Low-Velocity Impact Numerical Modelling. *Materials* **2018**, *11*, 1082. [CrossRef] [PubMed]
25. Pavlovic, A.; Fragassa, C.; Minak, G. Buckling Analysis of Telescopic Boom: Theoretical and Numerical Verification of Sliding Pads. *Tehnicki Vjesn.* **2017**, *24*, 729–735. [CrossRef]
26. Boria, S.; Pavlovic, A.; Fragassa, C.; Santulli, C. Modeling of Falling Weight Impact Behavior of Hybrid Basalt/Flax Vinylester Composites. *Procedia Eng.* **2016**, *167*, 223–230. [CrossRef]
27. Enderlein, M.; Ricoeur, A.; Kuna, M. Finite element techniques for dynamic crack analysis in piezoelectrics. *Int. J. Fract.* **2005**, *134*, 191–208. [CrossRef]
28. Chen, Y.M. Numerical computation of dynamic stress intensity factors by a Lagrangian finite-difference method (the HEMP code). *Eng. Fract. Mech.* **1975**, *7*, 653–660. [CrossRef]
29. Garcıa-Sanchez, F.; Zhang, C.; Saez, A. A two-dimensional time-domain boundary element method for dynamic crack problems in anisotropic solids. *Eng. Fract. Mech.* **2008**, *75*, 1412–1430. [CrossRef]
30. Chen, C.S.; Chen, C.H.; Pan, E. Three-dimensional stress intensity factors of a central square crack in a transversely isotropic cuboid with arbitrary material orientations. *Eng. Anal. Bound. Elem.* **2009**, *33*, 128–136. [CrossRef]

© 2018 by the authors. Licensee MDPI, Basel, Switzerland. This article is an open access article distributed under the terms and conditions of the Creative Commons Attribution (CC BY) license (http://creativecommons.org/licenses/by/4.0/).

Article

# On Air-Cavity Formation during Water Entry of Flexible Wedges

Riccardo Panciroli [1,*], Tiziano Pagliaroli [1] and Giangiacomo Minak [2]

1. Niccolò Cusano University, Engineering Faculty, via don Carlo Gnocchi 3, 00166 Rome, Italy; tiziano.pagliaroli@unicusano.it
2. Alma Mater Studiorum—Università di Bologna, DIN, via Fontanelle 40, 47121 Forlì, Italy; giangiacomo.minak@unibo.it
* Correspondence: riccardo.panciroli@unicusano.it

Received: 18 October 2018; Accepted: 3 December 2018; Published: 12 December 2018

**Abstract:** Elastic bodies entering water might experience fluid–structure interaction phenomena introduced by the mutual interaction between structural deformation and fluid motion. Cavity formation, often misleadingly named cavitation, is one of these. This work presents the results of an experimental investigation on the water entry of deformable wedges impacting a quiescent water surface with pure vertical velocity in free fall. The experimental campaign is conducted on flexible wedges parametrically varying the flexural stiffness, deadrise angle, and drop height. It is found that, under given experimental conditions, cavity pockets form beneath the wedge. Their generation mechanism might be ascribed to a differential between structural and fluid velocities, which is introduced by structural vibrations. Results show that the impact force during water entry of stiff wedges are always opposing gravity, while, in case flexible wedges temporarily reverse their direction, with the body that is being sucked into the water within the time frame between the cavity formation and its collapse. Severe impact might also generate a series of cavity generation and collapses.

**Keywords:** water entry; hydroelasticity; cavitation; FSI; SPH; slamming

---

## 1. Introduction

The impulsive nature of the hydrodynamic loading experienced by structures impacting the water might induce mechanical vibrations [1–3]. These introduce a series of so-called fluid–structure interaction (FSI) phenomena, such as air inclusions [4], ventilation, and cavitation [5], which are encountered in a wide range of water-entry problems, from naval [6–9] to aerospace applications [10–12]. In the literature, many experimental works investigated the hydrodynamic pressure at the fluid–structure interface during the water entry of rigid or very stiff bodies, showing that established analytical formulations [13,14] can be used with very high confidence [15]. However, in the case of FSI, theoretical formulations capable of predicting the hydrodynamic impact load developed for the water entry of rigid bodies become inaccurate. Due to the mutual coupling between the fluid motion and the structural deformation, the hydrodynamic loads that elastic bodies are subjected during the water entry might differ from the loads acting on rigid bodies [16]. The evolution of the wetted body area in time is an important characteristic of the impact, and variations of the structural shape due to its flexibility might affect the loads [17]. Such problems are still difficult to analyze and compute. Predicting structural deformations and stresses during the water entry of flexible structures is a major challenge, and a deep understanding of these FSI phenomena is direly needed. Most of the analytical and numerical works found in the literature [18–21] do not account for such FSI phenomena, since these can be neglected in cases where structural deformations are small, and hydrodynamic pressure is similar to the one experienced by a rigid body (see, e.g., References [22–27]).

The occurrence of cavitation during the water entry of flexible bodies has been predicted in the literature (e.g., References [7,28]). Hydroelasticity might facilitate cavitation [6,29], since pressure becomes subatmospheric during the second half of the first wet natural oscillation period. Reinhard [30] analytically predicted that there are conditions for which a wedge entering the water in free fall might form a cavity localized at its apex. Therein, the authors also mention that structure elasticity might enhance such a phenomenon. In the literature, the term cavitation is often misused, as many works relative to water-entry problems use this term to define the generation of cavity formation, rather than effective cavitation in its original meaning [31,32]. Korobkin [28] predicted that, in blunt bodies subjected to a sudden velocity drop, the liquid may separate from the entering body surface with the formation of a cavity. He defined this phenomenon as interface cavitation. In his model, which is based on Wagner's [13] theory, cavitation is supposed to happen when local pressure goes below the atmospheric. During the water entry of flat-bottom bodies, or geometries presenting a low deadrise angle, air can be trapped between the fluid and the structure during the early stage of the impact [33–37]. In such an occurrence, an air cushion is entrapped below the structure, lowering the hydrodynamic loading. Furthermore, the presence of entrapped air in the fluid is supposed to inhibit cavitation.

In this work, we report on experimental evidence about cavity formation during the water entry of flexible structures. Many experimental campaigns on the water impact of compliant bodies can be found in the literature [38–41], but to the authors' best knowledge, none of these reported on cavity formation for flexible wedges. The following sections report the most important experimental findings. First, we give insight on the analytical prediction of cavitation during the water entry of rigid bodies, showing the actual possibility of such effect. We then present the experimental setup, followed by details about the experimental results. The effects of cavity formation on impact dynamics are presented hereafter.

## 2. Cavitation Onset in Rigid Bodies' Water Entry as Predicted by Analytical Formulations

The dynamics of the water entry of rigid bodies can be accurately predicted by utilizing Wagner's model [13]. Such a solution relies on the concept of added mass (or virtual mass), where an increasing mass $m$ of water is considered to move with the body as it penetrates the water. In this framework, the velocity and acceleration of the impacting wedge are predicted by Wagner's model as [13]:

$$\dot{\xi} = \frac{MV_0}{M+m} = \frac{V_0}{1+\frac{\pi}{2}\rho\frac{(\pi/2)^2\xi^2}{M\tan^2(\beta)}} \qquad \ddot{\xi} = \frac{d}{dt}\frac{MV_0}{M+m} = -\frac{\pi\rho(\pi/2)^2}{MV_0\tan^2(\beta)}\xi\dot{\xi}^3, \qquad (1)$$

$\xi$ being the entry depth, $M$ the mass of the wedge per unit depth, $V_0$ the initial entry velocity, and $\beta$ the deadrise angle. Figure 1 shows a sketch of the problem. Wagner's model further allows to compute the pressure distribution along the wet portion of the body as:

$$\frac{p}{\rho} = \ddot{\xi}\sqrt{r^2-x^2} + \frac{\pi}{2}\frac{\dot{\xi}^2 r}{\tan(\beta)\sqrt{r^2-x^2}} - \frac{1}{2}\frac{\dot{\xi}^2 x^2}{r^2-x^2}, \qquad (2)$$

where $r$ is the horizontal projection of the wet length of the wedge, which reads $\frac{\pi}{2}\frac{\xi}{\tan\beta}$. Factor $\frac{\pi}{2}$ accounts for the water pile-up along the wetting edge due to the displaced water. Such value has been later found not to be a constant, as it actually varies with the deadrise angle [42]. Here, we utilize $\frac{\pi}{2}$ as this does not qualitatively alter the solution.

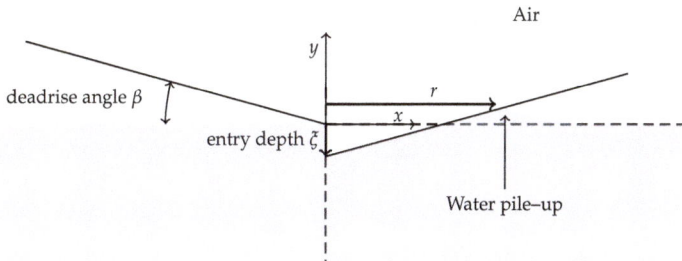

**Figure 1.** Sketch of the problem of the water entry of a rigid wedge. The wedge enters the water surface at $t = 0$ (**left**) and penetrates the water by an entry depth $\xi$ as time advances (**right**).

In case of water entry at a constant speed, the solution is self-similar in time. Otherwise, the component related to the acceleration may overcome the component associated to the velocity in Equation (2), with pressure eventually going subatmospheric at some locations.

Figure 2 shows an example Wagner's pressure-distribution prediction at several impact times. The location of peak pressure is constant, and the minimum pressure is always located at the keel of the wedge ($x = 0$), and equals

$$p_{\text{keel}} = \frac{1}{2}\rho\dot{\xi}^2 \pi \cot(\beta) + \ddot{\xi}\rho r. \tag{3}$$

We can therefore express the cavitation-onset condition in terms of dynamics components as

$$\dot{\xi}^2 + \ddot{\xi}\xi < (p_v - p_a)\frac{2\tan(\beta)}{\pi\rho}, \tag{4}$$

$p_a$ being atmospheric pressure and $p_v$ the vapor pressure. Using Equation (1), the left-hand side of the formula can only be written in terms of $\xi$ and equals

$$4\frac{\tan^4\beta\, M^2\, V_0^2\left(-\pi\rho\gamma^2\xi^2 + 2M\tan^2\beta\right)}{\left(2M\tan^2\beta + \pi\rho\gamma^2\xi^2\right)^3}, \tag{5}$$

showing a minimum at entry depth $\xi^* = 4\frac{\sqrt{\pi\rho M}\tan(\beta)}{\pi^2\rho}$ which, substituted in Equation (4), predicts that the minimum pressure at the keel is

$$\left.\dot{\xi}^2 + \ddot{\xi}\xi\right|_{\xi^*} = -\frac{1}{27}V_0^2. \tag{6}$$

It is noticeable that penetration depth $\xi^*$ might not be reached during impact due to insufficient initial-entry velocity. Wagner's solution thus predicts that cavitation during the water entry of a rigid wedge occurs if:

$$-\frac{1}{27}V_0^2 < (p_v - p_a)\frac{\tan(\beta)}{\frac{\pi}{2}\rho}. \tag{7}$$

Notably, $\xi^*$ is independent from impact velocity but is only a function of the geometrical data. Figure 3 (left) shows the normalized pressures versus the entry depth for various deadrise angles.

If we concentrate on the minimum value of the normalized pressure, we see that this increases with the deadrise angle, as displayed in Figure 3 (right). Here, only deadrise angles higher than 10° are shown, as lower values are known to lead to the so-called air-trapping phenomenon [4], that is, air is entrapped between the structure and the fluid during water entry to form an air cushion, decreasing the impact load and inhibiting cavitation.

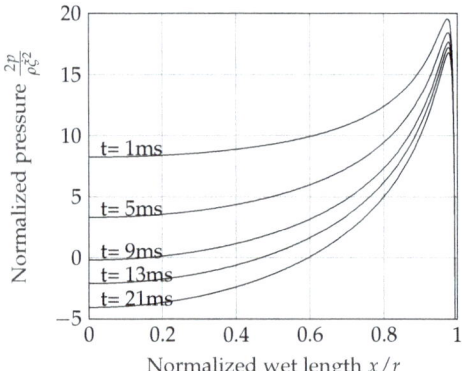

**Figure 2.** Wagner's predicted normalized pressure as a function of normalized wet length $x/r$ at several instants. The solution was calculated for a wedge weighing 5 kg/m, with a 20° deadrise angle, entering the water at 2 m/s with pure vertical velocity.

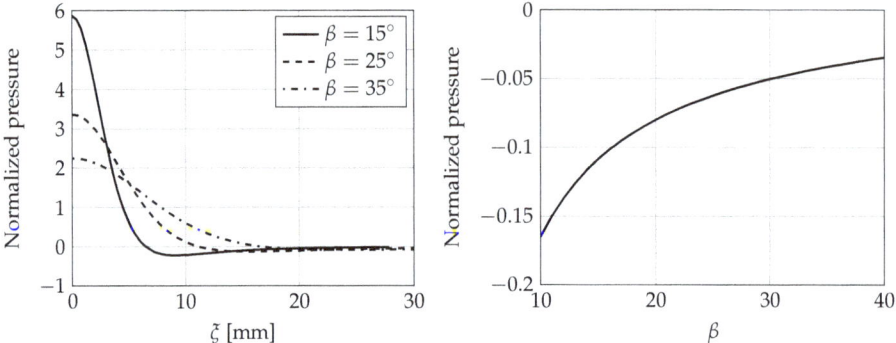

**Figure 3.** (**Left**) Normalized pressure versus entry depth for varying deadrise angles. (**Right**) Normalized minimum pressure versus deadrise angle.

## 3. Preliminary Experimental Evidences

The analytical solution presented in the previous section shows that cavitation (intended in its original meaning) might appear during the water entry of rigid bodies. However, such occurrence is extremely difficult to be attained as, to reach pressure lower than the vapor pressure, we need the combination of very high velocity and extremely lightweight bodies. Similar results were found by Reinhard in [30] for different body shapes. Further, all these solutions are 2D approximations and do not take into account the effects at the front and rear faces of the wedge (but are indeed valid for axial-symmetric bodies). We now discuss such issues by only referring to the keel edge, which is the location where minimum pressure is attained. The keel is always wet, being the first portion of the wedge touching the water. However, as the wedge enter the water, it digs a hole into it, pushing the water sideways. The front and back sides of the wedge actually remain dry and always "see" atmospheric pressure. The hole in the water will eventually collapse, but it takes a time way longer than the impact duration. Therefore, pressure at the keel varies from the theoretical one at its midspan to the atmospheric one at its vertexes. As the pressure at the keel goes below atmospheric, its effect forces some air to enter from the sides, forming a cavity beneath the wedge.

We performed some experiments on rigid wedges on this. In our experience, such an occurrence never happens in free-fall impact, but was indeed found when the deceleration of the wedge was imparted mechanically through a pneumatic actuator, or by a mechanic end-run forcing the wedge to suddenly decelerate. A schematic of the experimental apparatus with a mechanical stop is presented in Figure 4. Therein, the falling sledge holding the rigid wedge is forced to stop its motion by an end-run cable. The end-run cable is necessary to prevent the wedge to hit the water-tank floor. Cavity formation was observed as soon as the end-run cable stopped the wedge motion.

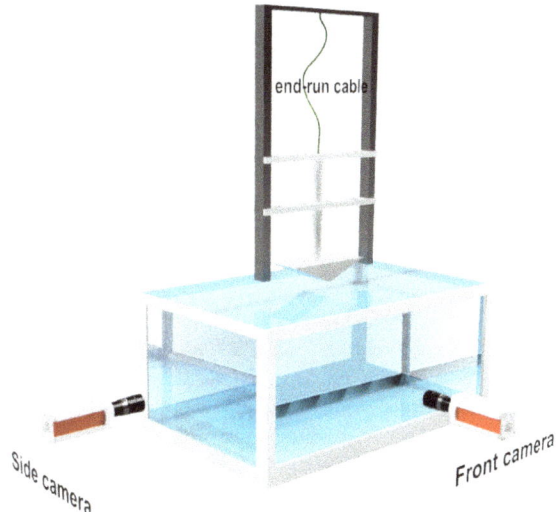

**Figure 4.** Schematics of a water-entry apparatus. The two possible locations (side and front) of the camera are showed. The sketch highlights the end-run cable used to stop the motion.

Figure 5 shows the evolution of a cavity formation in the neighborhood of the vertex at the keel of the wedge. The images were acquired through a phantom camera capturing the event from the side. The cavity-formation mechanisms is equally generating on both the wedge vertexes and only one of these is reported in the images for convenience. The images are presented here with the sole intent of explaining the mechanism of cavity formation during water entry, as these do not refer to the present experimental campaign, since the experimental apparatus utilized in this work does not allow to capture images from the side, only from the front side.

Results about cavity formation during the water entry of rigid body are to be considered artificial and are not presented here. We therefore never encountered cavity formation in the free-fall water entry of rigid bodies, whereas we indeed found it in the case of free-fall impact of flexible bodies. The following sections introduce the most important results about cavity formation during the water entry of flexible wedges.

**Figure 5.** Details of the vertex of the keel at several time instants, presenting the evolution of the cavity formation. Images are taken from the side.

## 4. Experimental Setup

Experiments were conducted on a drop-weight apparatus appositely assembled for the experimental campaign. The falling body was comprised of a sledge holding two panels joined together by a tunable hinge to form a V-shaped object. Deadrise angle $\beta$ may range smoothly from $0°$ to $50°$. The hinge held the panels in a cantilever configuration in a clamped-free boundary condition. Experiments were conducted on bodies with a sufficiently large deadrise angle ($>10°$) to avoid air-bubble inclusion [4].

The experimental setup was the same utilized by Panciroli [43]. Impact acceleration was measured by a Microstrain (Williston, VT) V-Link wireless accelerometer ($\pm 100$ g) located at the tip of the wedge. All reported accelerations are referenced to 0 g for the free-falling phase. Sampling frequency was set to its maximum of 4 kHz. Entering velocity was recorded by a laser sensor ILS 1402 ($\mu\epsilon$, Ortenburg, Germany) capturing the sledge position over 350 mm of ride at a frequency of 1.5 kHz with a resolution of 0.2 mm. Entry velocity was obtained through the central difference of the position signal.

The dynamics of the impact was captured through a high-speed camera looking through a window on the water tank. As mentioned before, only frontal views could be captured through the high-speed camera in the present experimental apparatus. Capturing frequency was set to 4 kHz with a definition of 1280 × 800 px. A vertical clear screen was placed inside the water tank just before the wedge preventing fluid spraying in the axial direction. The clearance between the screen and the falling body was about 2 mm, but it slightly varied during water entry due to varying fluid pressure moving and deforming the screen. The use of the screen was necessary to see the evolution of the fluid jet (pile-up) rising on the sides of the wedge, and to prevent fluid spraying through the camera that would affect the visibility of the image.

### 4.1. Specimens

Hydroelastic effects are influenced by the ratio between wetting time and the panel's lower natural frequency [6,44]. To vary the fundamental natural frequency of the panels, different stiffness to area-density ratios were utilized: Aluminum (A), E-glass (mat)/vinylester (V), and E-Glass (woven)/epoxy (W) 2 and 4 mm thick were used. All wedges were made of two panels 300 mm long and 250 mm wide. Aluminum and composite-panel material properties are listed in Table 1. Composite panels were produced by VARTM (vacuum assisted resin transfer molding) by infusion of vinylester resin on an E-Glass fiber mat, while the E-glass (woven $0°/90°$)/epoxy panels were produced in autoclave. The first three dry vibration frequencies of the panels are listed in Table 2.

Table 1. Material properties.

| Material | Abbr. | Elastic Moduli $E_1 = E_2$ | Poisson Ratio $\nu_{12}$ | Density $\rho$ (kg/m$^3$) |
|---|---|---|---|---|
| 6068 T6 | A | 68.0 GPa | 0.32 | 2700 |
| E-Glass/Vinylester | V | 20.4 GPa | 0.28 | 2050 |
| E-Glass/Epoxy | W | 30.3 GPa | 0.28 | 2015 |

All panels were equipped with two strain gauges per side, located at 25 and 120 mm from the reinforced tip. The reinforced tip was 27 mm long and used to connect the two panels to the aluminum sledge.

A very high number of drop tests were conducted [45,46], but high-speed imaging was performed on selected configurations only at the end of the experimental campaign, when the camera was rented for the purpose. Here, we therefore report on a limited number of experiments that are considered to be sufficient to qualitatively describe the effect of cavity formation on impact dynamics.

**Table 2.** First three theoretical dry natural frequencies of the panels composing the wedges.

| Abbr. | Material | Thickness | $\omega_1$ (Hz) | $\omega_2$ (Hz) | $\omega_3$ (Hz) |
|---|---|---|---|---|---|
| A2 | Aluminum | 2.0 mm | 18.0 | 112.8 | 316.1 |
| A4 | Aluminum | 4.0 mm | 36.0 | 225.7 | 632.2 |
| V2 | Fiberglass | 2.0 mm | 9.7 | 61.2 | 171.4 |
| V4 | Fiberglass | 4.0 mm | 19.7 | 123.6 | 346.2 |
| W2 | Fiberglass | 2.2 mm | 19.6 | 123.4 | 345.5 |
| W4 | Fiberglass | 4.4 mm | 37.8 | 236.9 | 663.4 |

## 5. Experimental Results

### 5.1. Wedge Deformation during Water Entry

In a previous study by the authors [44], it was shown that hydroelastic effects in the present experiments are ruled by a parameter $R$, which is proportional to the ratio between impact time and the first structural natural period. Results showed that the maximum impact-induced stress decreased if compared to the theoretical quasistatic solution when $R$ was lower than 100. Above this value, the structural response could be accurately predicted by a quasistatic approach. However, for values of $R$ lower than 100, interesting fluid–structure interaction phenomena might appear, such as ventilation and cavity formation.

Experiments on elastic wedges with a high deadrise angle entering the water at low velocity ($R > 100$) show that the panels initially slightly deform downward (thus showing a convex shape), to later deform upward due to hydrodynamic pressure.

This initial convex deformation has to be attributed to the effect of inertia: the hydrodynamic load acts on the wedge apex at first, leading the free edge to deform downward. As the wedge enters the water, the hydrodynamic load covers a larger area, and deformation due to pressure exceeds the deformation due to inertia, leading the panel to deform upward. As an example, Figure 6 shows deformation in time of a flexible wedge ($\beta = 30°$) entering the water at 4.2 m/s.

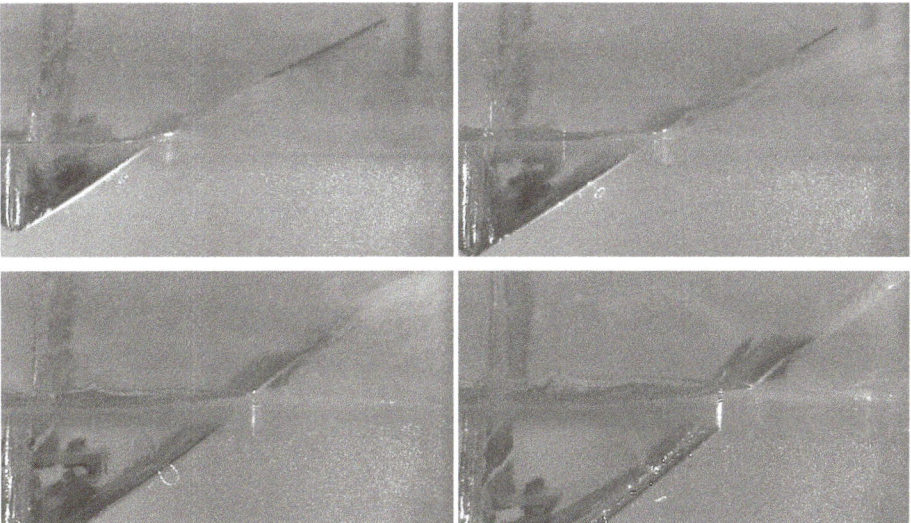

**Figure 6.** Deformation over time of a fiberglass/vinylester wedge with a deadrise angle of 30° entering the water at 4.2 m/s. Time advances left to right, top to bottom.

The strains at the vertex and at the middle of the panel assume a shape similar to those presented in Figure 7, which shows the example of a fiberglass/polyester wedge 4 mm thick with deadrise angle of 35° impacting from an impact height of 1.5 m. Generally speaking, in all these cases the panels are deforming downward for a very short portion of the total impact time, and the maximum positive strain (hence the stress) is way lower than the maximum negative strain reached later, revealing that, in case of "soft" impact, the influence of plate inertia is negligible.

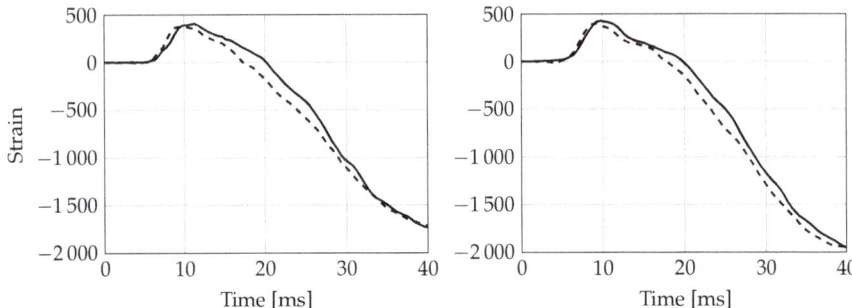

**Figure 7.** Signal recorded by two strain gauges during the water entry of a fiberglass/polyester wedge 4 mm thick with a deadrise angle of 35° entering the water from an impact height of 1.5 m (approximately 5 m/s). The graph on the right shows the strain measured at the center of the panel, while the graph on the left shows the strain close to the wedge tip. The full and the dashed lines represent two repetitions of the same experiment.

Conversely, moving to a stronger impact ($R < 100$, by lowering deadrise angle and panel stiffness, and increasing the impact velocity), the dynamic response showed very different results. Figure 8 shows the deformation in time of a flexible wedge entering the water from an impact height of 2.5 m. Due to the flexibility of the wedge and the very severe impact load, the panel was largely deforming downward at the beginning of impact. At its maximum deformation (top-right subfigure) the panel was almost horizontal at its free edge.

**Figure 8.** Deformation over time of wedge (V) 2 mm thick ($\beta = 20°$) entering the water at 6.7 m/s. Time advances left to right, top to bottom.

The overall deformation of the panels can be better caught by looking at the recordings of the strain gauges, reported in Figure 9. The two graphs show that the strain acquired by the gauge close to the wedge tip (left graph) is always negative, indicating a local convex deformation, while the strain gauge at the middle of the panel (right graph) is initially positive (suggesting that the deformation is locally concave), thus leading the panel free edge to deform downward at first. Note that, in this case, the maximum tensile stress, which is ruled by inertia, is in the same order of magnitude of the compressive stress, which is ruled by hydrodynamic load. It is thus necessary to consider the effect of inertia in the initial stages of the impact.

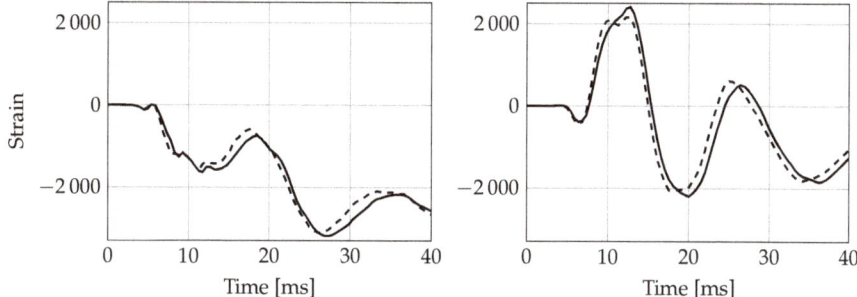

**Figure 9.** Signal acquired by two strain gauges during the water entry. The graph on the right shows the strain measured at the center of the panel, while the graph on the left shows the strain close to the wedge tip. Full and dashed lines are two repetitions of the same experiment.

### 5.2. Evidence of Cavity Formation from High-Speed Images

It was mentioned before that increasing the severity of the impact increases panel deformation. The high-speed images captured during the impact showed that the increasing deformation serves as onset for the generation of cavities within the liquid. It was found that in the most severe impacts, after maximum concave deformation is reached, a clearly visible cylindrical front of cavitating fluid is formed. Figures 10–13 show some examples of this phenomenon. Please note that the cylindrical cavity does not extend along the entire width of the wedge, but is only concentrated at the front and back edges, as shown in Section 3.

**Figure 10.** Evolution of the water entry of a wedge (W) ($\beta = 30°$) entering the water at $\approx 4.2$ m/s at 25, 40, and 55 ms from the impact.

**Figure 11.** Evolution of the water entry of a wedge (W) ($\beta = 15°$) entering the water at $\approx 6.7$ m/s at 16.6, 20, and 23.3 ms from the impact. The arrow highlights the cylindrical front of the cavitating area.

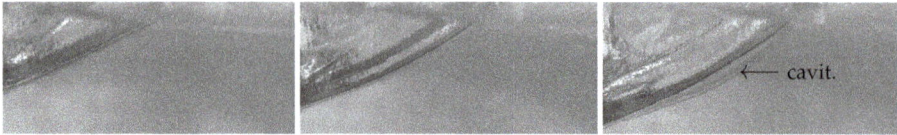

**Figure 12.** Evolution of the water entry of a wedge (V) ($\beta = 20°$) entering the water at ≈4.3 m/s at 40, 53.3, and 66.6 ms from the impact. The arrow highlights the cylindrical front of the cavitating area.

**Figure 13.** Evolution of the water entry of a wedge (V) ($\beta = 20°$) entering the water at ≈6 m/s at 40, 53.3 , and 66.6 ms from the impact. The arrow highlights the cylindrical front of the cavitating area.

Figure 10 is taken as reference. It shows a wedge with a 30° deadrise angle entering the water at 4.3 m/s. In this case, the deformation of the panels is very low, and all the pictures showed a smooth and uniformly colored water region. As the severity of the impact increased, it was found that a fluid region with a cylindrical waveform front was generated in the fluid (right pictures in Figures 12 and 13). Such a phenomenon is always found developing after maximum concave deformation is attained (central pictures). We comment that the cavity shape is not in line with the analytical predictions, as it does not seem to originate at the keel, which is the location of the minimum pressure. The cavity eventually collapses during water entry. In the most severe impact cases, successive pockets (with decreasing amplitude) might generate and collapse.

Figure 14 shows that, for a given wedge (V) ($\beta = 20°$), the maximum dimension of the wavefront increases with impact velocity, as does maximum deformation. Please note that the images in Figure 14 correspond to the same impact time. In fact, as mentioned before, the cavity-formation phenomena initiate after the maximum deformation of the wedge is reached; such deformation is ruled by the first natural frequency of the panel, which is not influenced by entry velocity.

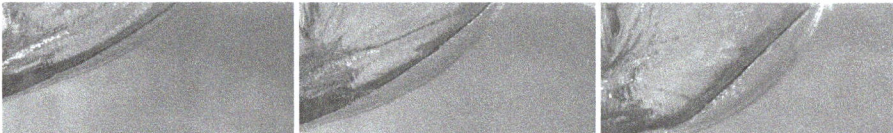

**Figure 14.** Maximum dimension of the wavefront during the water entry of a wedge (V) ($\beta = 20°$) entering the water at 4.3, 6, and 6.7 m/s.

It was previously shown that the cavity cross-sectional area increases with impact velocity and structural compliance. Here, we provide insight on its effect on the acceleration of the body during the impact.

Figure 15 shows the time traces of acceleration recorded during the water entry of a composite wedge (V2) with $\beta = 20°$ entering the water at 5.6 m/s. Three repetitions of the same experiment are shown (solid, dashed, and dotted lines). Images from the high-speed camera at characteristic time instants of the impact were superimposed to the graph to get a better overview of the impact dynamics. Notably, acceleration suddenly turns negative at about 20 ms from the impact (positive acceleration in the plot is opposite to gravity, and free fall is referenced as 0). Such behavior is very uncommon in water entry, as it represents a body being sucked into the water. The analysis of the high-speed images showed that acceleration starts its decreasing trend to eventually attain negative values after the cavity is generated within the fluid. It was further found that acceleration turns positive as the cavity pockets collapse. Results thus evidence strong relation between cavity evolution and impact dynamics. The lower graph in Figure 15 further shows the time trace of the strain during the impact. Strain signals are acquired by two strain gauges located at 30 mm and 130 mm from the keel (which is almost at the mid-span). Results show that the negative acceleration initiates as the maximum concave deformation is attained, that is, when the strain gauge at the midspan reaches its peak, approximately 18 ms after impact. By comparing the images from the high-speed camera with the time trace of the strains, it appears that the cavity starts as the maximum concave deformation of the panel is reached.

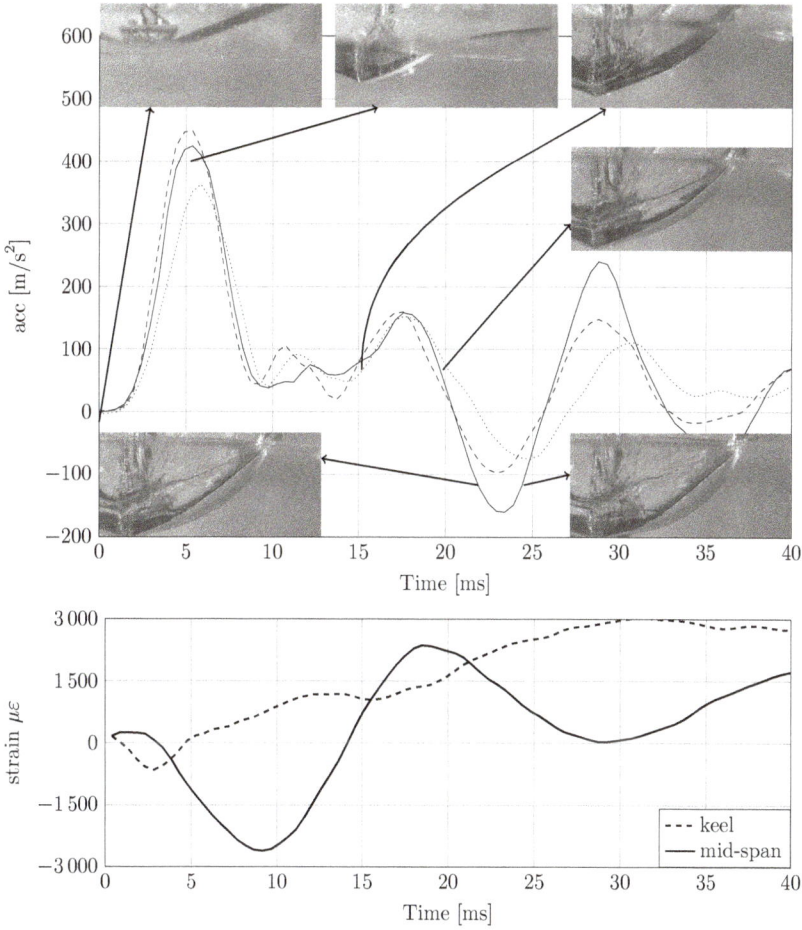

**Figure 15.** Graph of the recorded acceleration of a wedge (V) 2 mm thick ($\beta = 20°$) entering the water at 5.6 m/s, and high-speed camera images captured at 0, 5, 15, 20, 22.5, and 25 ms.

## 6. Conclusions

Experimental drop tests of flexible wedges were performed to study the fluid–structure interaction phenomena that generate during water entry, with particular attention to the cavity-formation process and its effect on impact dynamics. It was found that, when the deflection of the wedge is small, no fluid–structure interaction phenomena appear and established analytical formulations for rigid bodies can be used to evaluate the impact force and the hydrodynamic pressure. However, large structural deformations were found to have strong effects on the fluid motion. In particular, large deformations might introduce air pockets characterized by a cylindrical wavefront originating at the fluid/structure interface. Such a cavity is found to equally originate from the front and rear faces of the structure, and their shape is not in line with analytical predictions, as it seems to not originate from the keel, where pressure is supposed to be the lowest, but along the entire wet length of the structure. The cavity eventually collapses during water entry, and, in the most severe impact cases, successive pockets (with decreasing amplitude) might repetitively generate and collapse. The pockets are found to increase with the impact velocity and the flexibility of the wedge. The largest cavities were further found to strongly affect impact dynamics, up to the cases where the overall acceleration of the body turns negative within the timeframe between cavity formation and its collapse. The analysis of the panel's deformation suggests that the cavity initiates as the maximum concave deformation of the panel is reached, further confirming that cavity formation is led by structural vibrations.

**Author Contributions:** Conceptualization, G.M.; data curation, R.P.; funding acquisition, R.P. and G.M.; resources, T.P.; supervision, G.M.; visualization, T.P.; writing—original draft, R.P.; writing—review & editing, T.P.

**Funding:** This research received no external funding.

**Conflicts of Interest:** The authors declare no conflict of interest.

## References

1. Qin, Z.; Batra, R.C. Local slamming impact of sandwich composite hulls. *Int. J. Solids Struct.* **2009**, *46*, 2011–2035. [CrossRef]
2. Carcaterra, A.; Ciappi, E. Prediction of the Compressible Stage Slamming Force on Rigid and Elastic Systems Impacting on the Water Surface. *Nonlinear Dyn.* **2000**, *21*, 193–220. [CrossRef]
3. Carcaterra, A.; Ciappi, E. Hydrodynamic shock of elastic structures impacting on the water: Theory and experiments. *J. Sound Vib.* **2004**, *271*, 411–439. [CrossRef]
4. Panciroli, R.; Minak, G. Experimental evaluation of the air trapped during the water entry of flexible structures. *Acta Imeko* **2014**, *3*, 63–67. [CrossRef]
5. Kapsenberg, G.K. Slamming of ships: where are we now? *Philos. Trans. R. Soc. A Math. Phys. Eng. Sci.* **2011**, *369*, 2892–2919. [CrossRef] [PubMed]
6. Faltinsen, O.M. Hydroelastic slamming. *J. Mar. Sci. Technol.* **2000**, *5*, 49–65. [CrossRef]
7. Faltinsen, O.M.; Landrini, M.; Greco, M. Slamming in marine applications. *J. Eng. Math.* **2004**, *48*, 187–217. [CrossRef]
8. Fragassa, C. Engineering Design Driven by Models and Measures: The Case of a Rigid Inflatable Boat. *Preprints* **2018**. [CrossRef]
9. Fragassa, C.; Minak, G. Measuring Deformations in a Rigid-Hulled Inflatable Boat. *Key Eng. Mater.* **2017**, *754*, 295–298. [CrossRef]
10. Seddon, C.; Moatamedi, M. Review of water entry with applications to aerospace structures. *Int. J. Impact Eng.* **2006**, *32*, 1045–1067. [CrossRef]
11. Campbell, J.C.; Vignjevic, R. Simulating structural response to water impact. *Int. J. Impact Eng.* **2012**, *49*, 1–10. [CrossRef]
12. Hughes, K.; Vignjevic, R.; Campbell, J.; De Vuyst, T.; Djordjevic, N.; Papagiannis, L. From aerospace to offshore: Bridging the numerical simulation gaps–Simulation advancements for fluid structure interaction problems. *Int. J. Impact Eng.* **2013**, *61*, 48–63. [CrossRef]
13. Wagner, H. {Ü}ber Sto{ß}- und Gleitvorg{ä}nge an der Oberfl{ä}che von Fl{ü}ssigkeiten. *ZAMM Z. Angew. Math. Mech.* **1932**, *12*, 193–215. [CrossRef]
14. Chuang, S.L. *Investigation of Impact of Rigid and Elastic Bodies with Water*; NSRDC Report No. 3248; David Taylor Model Basin Reports; PN: Bethesheda, MA, USA, 1970.
15. Panciroli, R.; Porfiri, M. Evaluation of the pressure field on a rigid body entering a quiescent fluid through particle image velocimetry. *Exp. Fluids* **2013**, *54*, 1630. [CrossRef]
16. Korobkin, A.; Parau, E.I.; Vanden-Broeck, J.M. The mathematical challenges and modelling of hydroelasticity. *Philos. Trans. Ser. A Math. Phys. Eng. Sci.* **2011**, *369*, 2803–2812. [CrossRef] [PubMed]
17. Korobkin, A.; Guéret, R.; Malenica, Š. Hydroelastic coupling of beam finite element model with Wagner theory of water impact. *J. Fluids Struct.* **2006**, *22*, 493–504. [CrossRef]

18. Das, K.; Batra, R.C. Local water slamming impact on sandwich composite hulls. *J. Fluids Struct.* **2011**, *27*, 523–551. [CrossRef]
19. Zhao, R.; Faltinsen, O.; Aarsnes, J. Water entry of Arbitrary Two-Dimensional sections with and without flow separation. In *Twenty-First Symposium on Naval Hydrodynamics*; The National Academies Press: Washington, DC, USA, 1997.
20. Scolan, Y. Hydroelastic behaviour of a conical shell impacting on a quiescent-free surface of an incompressible liquid. *J. Sound Vib.* **2004**, *277*, 163–203. [CrossRef]
21. Wu, G.X.; Sun, H.; He, Y.S. Numerical simulation and experimental study of water entry of a wedge in free fall motion. *J. Fluids Struct.* **2004**, *19*, 277–289. [CrossRef]
22. Von Karman, T. *The Impact on Seaplane Floats, during Landing*; NACA-TN-321; National Advisory Committee for Aeronautics: Washington, DC, USA, 1929.
23. Backer, G.D.; Vantorre, M.; Beels, C.; Pré, J.D.; Victor, S.; Rouck, J.D.; Blommaert, C.; De Backer, G.; Vantorre, M.; Beels, C.; et al. Experimental investigation of water impact on axisymmetric bodies. *Appl. Ocean Res.* **2009**, *31*, 143–156. [CrossRef]
24. El Malki Alaoui, A.; Nême, A.; Tassin, A.; Jacques, N.; Alaoui, A.E.M.; Nême, A.; Tassin, A.; Jacques, N. Experimental study of coefficients during vertical water entry of axisymmetric rigid shapes at constant speeds. *Appl. Ocean Res.* **2012**, *37*, 183–197. [CrossRef]
25. Chuang, S.L.; Milne, D.T. *Drop Tests of Cone to Investigate the Three-Dimensional Effect Of Slamming*; NRDC Report No. 3543; Naval Ship Research and Development Center: Washington, DC, USA, 1971.
26. Jalalisendi, M.; Shams, A.; Panciroli, R.; Porfiri, M. Experimental reconstruction of three-dimensional hydrodynamic loading in water entry problems through particle image velocimetry. *Exp. Fluids* **2015**, *56*, 1–17. [CrossRef]
27. Jalalisendi, M.; Osma, S.J.; Porfiri, M. Three-dimensional water entry of a solid body: A particle image velocimetry study. *J. Fluids Struct.* **2015**, *59*, 85–102. [CrossRef]
28. Korobkin, A. Cavitation in liquid impact problems. In Proceedings of the Fifth International Symposium on Cavitation (CAV2003), Osaka, Japan, 1 January 2003; Volume 2, pp. 1–7.
29. Faltinsen, O.M. The effect of hydroelasticity on ship slamming. *Philos. Trans. R. Soc. A Math. Phys. Eng. Sci.* **1997**, *355*, 575–591. [CrossRef]
30. Reinhard, M.; Korobkin, A.A.; Cooker, M.J. Cavity formation on the surface of a body entering water with deceleration. *J. Eng. Math.* **2015**. [CrossRef]
31. Bivin, Y.K.; Glukhov, Y.M.; Permyakov, Y.V. Vertical entry of solids into water. *Fluid Dyn.* **1986**, *20*, 835–841. [CrossRef]
32. Yadong, W.; Xulong, Y.; Yuwen, Z. Natural Cavitation in High Speed Water Entry Process. In Proceedings of the 1st International Conference on Mechanical Engineering and Material Science, Shanghai, China, 28–30 December 2012; Atlantis Press: Paris, France, 2012; pp. 46–49. [CrossRef]
33. Korobkin, A.A.; Khabakhpasheva, T.I.; Wu, G.X. Coupled hydrodynamic and structural analysis of compressible jet impact onto elastic panels. *J. Fluids Struct.* **2008**, *24*, 1021–1041. [CrossRef]
34. Korobkin, A.; Ellis, A.S.; Smith, F.T. Trapping of air in impact between a body and shallow water. *J. Fluid Mech.* **2008**, *611*, 365–394. [CrossRef]
35. Hicks, P.D.; Ermanyuk, E.V.; Gavrilov, N.V.; Purvis, R.; Mechanics, F. Air trapping at impact of a rigid sphere onto a liquid. *J. Fluid Mech.* **2012**, *695*, 310–320. [CrossRef]
36. Cuomo, G.; Piscopia, R.; Allsop, W. Evaluation of wave impact loads on caisson breakwaters based on joint probability of impact maxima and rise times. *Coast. Eng.* **2011**, *58*, 9–27. [CrossRef]
37. Ma, Z.H.; Causon, D.M.; Qian, L.; Mingham, C.G.; Mai, T.; Greaves, D.; Raby, A. Pure and aerated water entry of a flat plate.. *Phys. Fluids* **2016**, *28*, 016104. [CrossRef]
38. Panciroli, R.; Porfiri, M. Hydroelastic impact of piezoelectric structures. *Int. J. Impact Eng.* **2014**, *66*, 18–27. [CrossRef]
39. Jalalisendi, M.; Porfiri, M. Water entry of compliant slender bodies: Theory and experiments. *Int. J. Mech. Sci.* **2018**, *149*, 514–529. [CrossRef]
40. Panciroli, R.; Porfiri, M. Analysis of hydroelastic slamming through particle image velocimetry. *J. Sound Vib.* **2015**, *347*, 63–78. [CrossRef]
41. Shams, A.; Zhao, S.; Porfiri, M. Water impact of syntactic foams. *Materials* **2017**, *10*. [CrossRef] [PubMed]

42. Mei, R.; Luo, L.S.; Shyy, W. An Accurate Curved Boundary Treatment in the Lattice Boltzmann Method. *J. Comput. Phys.* **1999**, *155*, 307–330. [CrossRef]
43. Panciroli, R.; Abrate, S.; Minak, G.; Zucchelli, A. Hydroelasticity in water-entry problems: Comparison between experimental and SPH results. *Compos. Struct.* **2012**, *94*, 532–539. [CrossRef]
44. Panciroli, R.; Abrate, S.; Minak, G. Dynamic response of flexible wedges entering the water. *Compos. Struct.* **2013**, *99*, 163–171. [CrossRef]
45. Panciroli, R. Hydroelastic Impacts of Deformable Wedges. Ph.D. Thesis, Alma Mater Studiorum Università di Bologna, Bologna, Italy, 2012.
46. Panciroli, R. Hydroelastic Impacts of Deformable Wedges. In *Solid Mechanics and Its Applications*; Abrate, S., Castanié, B., Rajapakse, Y.D.S., Eds.; Springer: Dordrecht, The Netherlands, 2013; Volume 192, pp. 1–45.

© 2018 by the authors. Licensee MDPI, Basel, Switzerland. This article is an open access article distributed under the terms and conditions of the Creative Commons Attribution (CC BY) license (http://creativecommons.org/licenses/by/4.0/).

Article

# Geometric Evaluation of Stiffened Steel Plates Subjected to Transverse Loading for Naval and Offshore Applications

João Pedro T. P. de Queiroz [1], Marcelo L. Cunha [1], Ana Pavlovic [2], Luiz Alberto O. Rocha [3], Elizaldo D. dos Santos [1], Grégori da S. Troina [1] and Liércio A. Isoldi [1,*]

1. Programa de Pós-Graduação em Engenharia Oceânica (PPGEO), Escola de Engenharia (EE), Universidade Federal do Rio Grande (FURG), Rio Grande 96203-900, Brazil; jopetpq@gmail.com (J.P.T.P.d.Q.); marcelolamcunha@hotmail.com (M.L.C.); elizaldosantos@furg.br (E.D.d.S.); gregori.troina@gmail.com (G.d.S.T.)
2. Dipartimento di Ingegneria Industriale, Alma Mater Studiorum Università di Bologna, Viale del Risorgimento 2, 40136 Bologna, Italy; ana.pavlovic@unibo.it
3. Programa de Pós-Graduação em Engenharia Mecânica, Universidade do Vale do Rio dos Sinos (UNISINOS), São Leopoldo 93020-190, Brazil; luizor@unisinos.br
* Correspondence: liercioisoldi@furg.br; Tel.: +55-53-3233-6916

Received: 16 November 2018; Accepted: 28 December 2018; Published: 7 January 2019

**Abstract:** This work searched for the optimal geometrical configuration of simply supported stiffened plates subjected to a transverse and uniformly distributed load. From a non-stiffened reference plate, different geometrical configurations of stiffened plates, with the same volume as the reference plate, were defined through the constructal design method. Thus, applying the exhaustive search technique and using the ANSYS software, the mechanical behaviors of all the suggested stiffened plates were compared to each other to find the geometrical configuration that provided the minimum deflection in the plate's center when subjected to this loading. The optimum geometrical configuration of stiffeners is presented at the end of this work, allowing a reduction of 98.57% for the central deflection of the stiffened plate if compared to the reference plate. Furthermore, power equations were adjusted to describe the deflections for each combination of longitudinal and transverse stiffeners as a function of the ratio between the height and the thickness of the stiffeners. Finally, a unique equation for determining the central deflections of the studied stiffened plates based only on the number of longitudinal stiffeners without significantly losing accuracy has been proposed.

**Keywords:** plate; stiffeners; constructal design; finite element method; deflection; numerical simulation

## 1. Introduction

Steel plates are structural components that are resistant to mechanical loads and used in several sectors of engineering—from the construction of bridges and buildings to airplanes and vessels. In the naval sector, for example, a great number of these structures are required to resist the extremely high loads that arise either because of the weight of their own structure or the sea conditions faced during navigation [1].

Due to the aforementioned high loadings, these structures must be extremely resistant, particularly to shear stresses and bending moments. Adding more material can increase their strength, although it also increases the weight of the structure, which should be avoided, especially when developing projects for the offshore [2] and naval [1] industries. In this context, the application of mechanical stiffeners arises. These stiffeners are long structural profiles fixed to the plates in the horizontal, transverse, or both horizontal and transverse directions. They are employed to improve the rigidity of the structural components that are made out of thin plates.

The structural analysis of stiffened plates has been the focus of countless studies. In [3], the restrictions method in statistical analysis was applied, via finite elements, to stiffened plates with concentric and eccentric stiffeners. Reference [4] studied stiffened plates under transversal loading using the sequential quadratic programming method, supposing the structure as a plate firmly connected to stiff beams and minimizing the total energy of the system. In [5] was utilized a methodology that considered the stresses and the deformations on the plate's plane, as well as the axial stresses and the deformations on the beams. These stresses were evaluated at the interface between the plates and the stiffeners, and the solution of the differential equations was obtained through the analog equation method (AEM).

More recently, reference [6] presented numerical studies regarding ribbed floor slabs, indicating that the consideration of eccentricity between the plate and the reinforcement beam results in a reduction in structure displacement. The steel or concrete slabs were simulated with the software ANSYS®, using shell elements (SHELL63) for modeling the plate and beam elements (BEAM44) for the stiffeners. Moreover, from [7], four computational models for stiffened plates using the finite element SHELL93 (triangular and quadrilateral) and SOLID95 (tetrahedral and hexahedral), also by means of the software ANSYS®, were suggested. All the models were verified by comparing their results with the ones obtained by other authors used as a reference.

The present article investigated different geometrical configurations that aimed at increasing the structural rigidity of stiffened plates without adding material that would result, consequently, in heavier structures. In doing so, a steel non-stiffened plate with a determined volume was adopted as a reference. Using the constructal design method, part of the steel volume was converted to stiffeners due to the parameter stiffeners volume fraction ($\varnothing$). The constructal design method was developed based on the constructal law, which states: "For a finite-size flow system to persist in time (to survive), its configuration must evolve in such a way that it provides easier access to the currents that flow through it" [8]. In practical terms, the constructal design method is based on objectives and restrictions [9]. In the geometrical evaluation of a given physical system, it is necessary and sufficient to define at least one objective (a performance parameter to be improved), the degrees of freedom (variables), and the geometrical restrictions (fixed parameters). The degrees of freedom are free to vary, but it should respect the imposed constraints [10].

It is noteworthy that the total volume of steel was kept constant and equal to the volume of the reference plate. Thereby, several different geometrical configurations of stiffened plates were established by means of varying the following degrees of freedom: number of longitudinal stiffeners ($N_{ls}$), number of transverse stiffeners ($N_{ts}$), and the ratio between the height and the thickness of the stiffeners ($h_s/t_s$). The mechanical behavior of the plates was numerically simulated, considering them as simply supported and subjected to a transverse and uniformly distributed load (pressure), aiming to reproduce a practical situation of a part of a ship deck or offshore oil platform deck. The software ANSYS Mechanical APDL®, which is based on the finite element method (FEM), was utilized to perform the simulations. Thus, the numerical results for the central deflection of the different proposed plates were compared to each other through the exhaustive search technique, allowing the geometrical configuration that would lead to a superior mechanical performance regarding the deflections to be determined.

## 2. Computational Modeling

Problems involving thin plates without stiffeners are difficult to solve because they involve a fourth order differential equation [11]. Concerning stiffened plates, an analytical solution becomes practically unreachable, especially when the intent of the research involves the evaluation of a great number of scenarios. Thus, to develop this work, it was decided to use the software ANSYS Mechanical APDL® (version 14.0, ESSS, Canonsburg, PA, USA), which applies the finite element method (FEM) to successfully solve several kinds of problems from structural [12] to heat transfer [13], and embraces all types of materials from composites [14–16] to concrete [17] and ceramics [18].

Two computational models were developed, one for the non-stiffened plate (used for simulating the reference plate, with no stiffeners) and the other for stiffened plates (used to simulate the geometrical configurations proposed by the constructal design method). In both models, the finite element SHELL93 with a regular quadrilateral mesh was adopted because it is the one indicated for problems involving thin plates and curved shells and it has produced suitable results in similar research [6,19,20]. The Sparse Matrix Direct Solver was employed in both computational models. This numerical method is based on a direct elimination of equations by means of the factorization of an initially very sparse linear system of equations into a lower triangular matrix followed by forward and backward substitution using this triangular system [21].

To ensure that the computational model is according to the reality that it is meant to simulate, it is possible to validate or verify it. The validation procedure was based on a comparison between the results obtained by the numerical model and laboratory experiments [7,10]. The verification procedure consisted of confronting the generated numerical results with the numerical results obtained by other authors and/or with values of analytical resolutions [11].

It was also important to perform the mesh convergence study, which determines the minimum refinement that provides a numerical result sufficiently accurate. Hence, numerical simulations with successive refinements were performed until there was no significant difference between the obtained results from meshes with successive refinements, defining, thus, the discretization that provides an independent numerical solution. Therefore, all the results achieved in this research were obtained using converged meshes, which exhibited a relative difference of less than 0.15% between the results of consecutive refinements according to the criteria established in this work.

## 2.1. Verification of the Computational Model for the Non-Stiffened Plate

Concerning the model of the non-stiffened plate, the verification process occurred through a comparison of its results with the ones acquired by numerical studies of [5,6] and with the analytical solution of [22]. In this case, the dimensions of the plate were defined according to Figure 1, while the material of the plate presented a Young's modulus ($E$) and Poisson's ratio ($\nu$) of 30 GPa and 0.154, respectively, characterizing a concrete slab. The boundary conditions were established as simple, supported edges and a pressure of 10 kPa was applied in the negative z-direction (see Figure 1). Figure 2 presents the results obtained by the references, as well as the computational results generated in the present work.

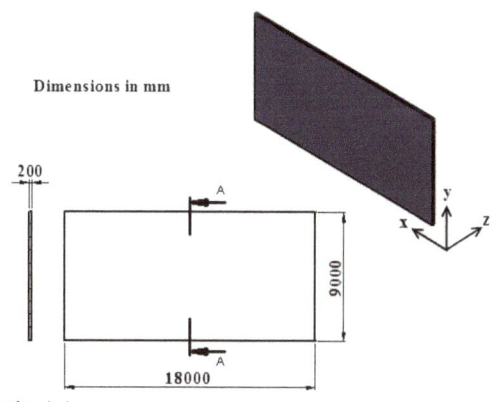

**Figure 1.** Non-stiffened plate used for computational model verification.

From Figure 2, it can be noticed that the values of the central deflection $w$ from the developed model converge closely with the ones attained by references [5,6,22] as the number of finite elements during the mesh convergence study increased. The present work presented differences of 2.50%, 1.86%, and 1.23% when compared with the results of [5,6,22], respectively. Therefore, it is possible to state that this model has been properly verified.

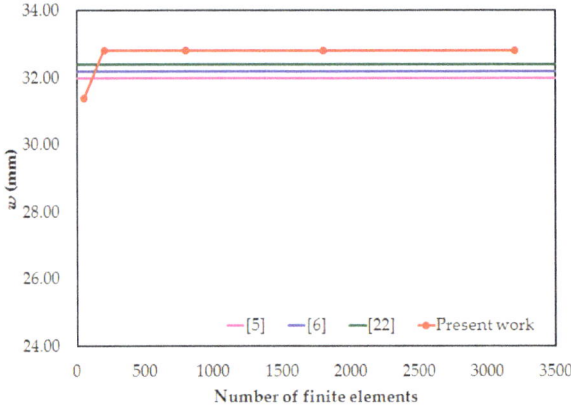

**Figure 2.** Non-stiffened plate computational model verification.

### 2.2. Verification of the Computational Model for the Stiffened Plate

Similarly, the model for the stiffened plates was verified by comparing its values of central deflection to the ones obtained by the numerical research executed by [3,4,6,7]. The stiffened plate was considered as simply supported, having a constant pressure of 68.95 kPa applied in the negative z-direction on its surface without stiffeners (see Figure 3). The material of the plate and the stiffeners was steel with $E$ and $\nu$ of 206.84 GPa and 0.3, respectively.

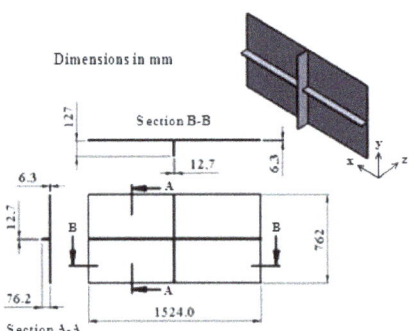

**Figure 3.** Stiffened plate used for computational model verification.

According to Figure 4, as a result of the mesh convergence study, the values achieved by the developed computational model converge narrowly with the ones found in [7]. However, they are not similar to the values found by the other authors. Differences of 26.95%, 40.26%, and 30.55% were found when comparing the results of the present work with those obtained by [3,4,6], respectively. As reported by [7], this difference is due to the fact that a less accurate finite element than the SHELL93 was employed, such as in [6], which used the SHELL63 finite element for the plate. Besides, the mesh used by these older studies [3,4,6] was likely too coarse. However, when the numerical result of the present study was compared with the result of [7], which was obtained with a more accurate 3D SOLID95 finite element, a difference of only 2.52% was achieved. Thus, for the reasons explained here, it can be considered that the computational model developed for stiffened plates has been properly verified, despite the divergence found with the studies performed by [3,4,6].

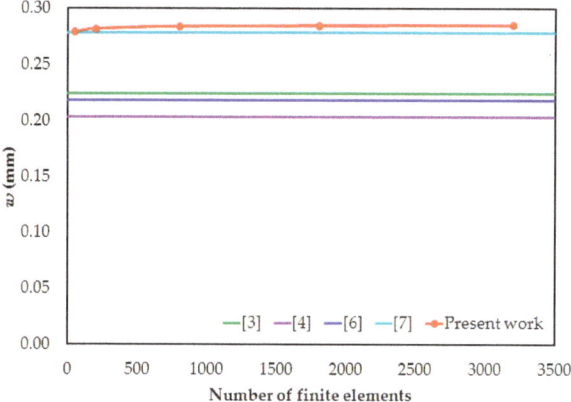

Figure 4. Stiffened plate computational model verification.

## 3. Constructal Design Method

In the present work, the constructal design method application enabled defining a set of stiffened plates with various geometric configurations that were numerically simulated and compared through the exhaustive search technique. To do so, the dimensions of a non-stiffened plate to be used as a reference were defined: length $a$ = 2000 mm, width $b$ = 1000 mm, and thickness $t$ = 20 mm.

All geometric configurations of the stiffened plates were formed from the reference plate. The stiffened plate had the same in-plane dimensions as the reference plate, i.e., the length $a$ and the width $b$ were kept constant. However, a portion of the reference plate material was transformed into stiffeners by reducing its thickness $t$. The ratio between the volume of the reference plate and the volume of material employed as stiffeners, called stiffeners volume fraction, is defined by:

$$\phi = \frac{V_s}{V_r} = \frac{N_{ls}(ah_st_s) + N_{ts}[(b - N_{ls}t_s)h_st_s]}{abt} \quad (1)$$

where $V_s$ is the volume of the reference plate transformed into stiffeners; $V_r$ is the total volume of the plate used as reference; $N_{ls}$ and $N_{ts}$ represent, respectively, the number of longitudinal and transverse stiffeners; $h_s$ and $t_s$ are, respectively, the height and thickness of the stiffeners; and $a$, $b$, and $t_p$ are, respectively, the length, width, and thickness of the stiffened plate, as indicated in Figure 5.

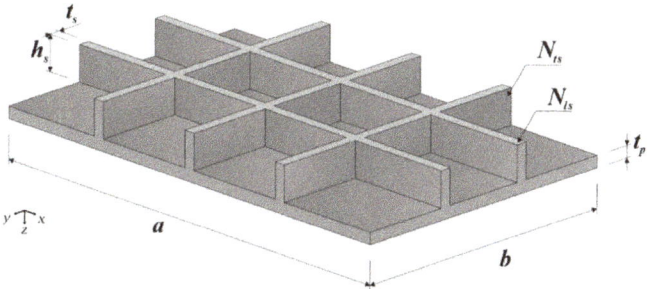

Figure 5. Stiffened plate with two longitudinal and three transverse stiffeners.

In order to accomplish the objective of finding a stiffened plate that presents a minor central deflection, the evaluated degrees of freedom were $N_{ls}$, $N_{ts}$, and $h_s/t_s$; while the imposed restrictions were that all stiffened plates have the same material volume, length, and width as the reference plate. Furthermore, since the present work adopted only one value of $\emptyset$ (which was defined as 0.4), it became an additional geometrical restriction of the analysis. In other words, 40% of the volume of the reference plate was turned into stiffeners of different configurations and geometries, due to the variation of the degrees of freedom $N_{ls}$, $N_{ts}$, and $h_s/t_s$. In addition, values of $N_{ls}$ and $N_{ts}$ were adopted between 2 to 5; other restrictions due to geometrical limitations of the problem included: $h_s \leq 300$ mm, avoiding a disproportionality between the height of the stiffeners and the width of the plate; and $h_s/t_s \geq 1$, in order to avoid obtaining stiffeners with heights greater than their thickness, which would over reduce the moment of inertia of these reinforcements. The thickness of the stiffeners was adopted according to standard sizes of steel plates, varying from 3.75 mm (1/8 in) to 75.2 mm (3 in). Furthermore, in all simulations, the applied load conditions were the same—that is, a uniformly distributed transverse loading (pressure) of 10 kPa—aiming to guarantee that all cases studied had a linear-elastic behavior. Similarly, regarding the boundary conditions, all simulated plates were simply supported on all four of their edges. Moreover, the steel used in these structures had a Young's modulus and Poisson's ratio of 200 GPa and 0.3, respectively.

Figure 6 depicts the methodological structure used in the definition of the geometrical configurations of the stiffened plates that composed the analyzed search space.

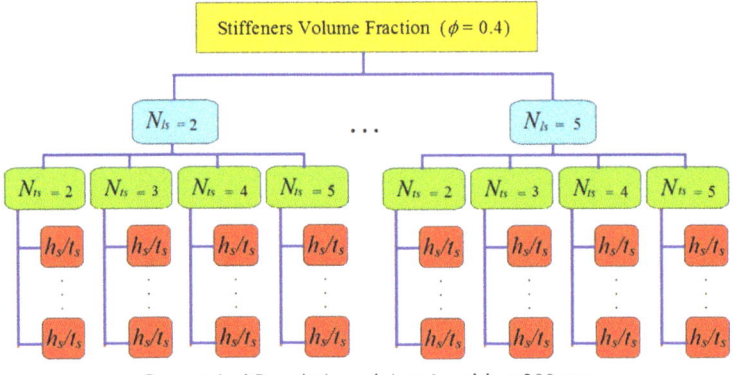

**Figure 6.** Application of the constructal design method in the definition of the search space.

## 4. Results and Discussion

Following the scheme presented in Figure 6, various numerical simulations of the stiffened plates were performed (a total of 166 simulations).

The non-stiffened reference plate was numerically simulated using a converged regular mesh with 100 mm quadrilateral shaped elements; a central displacement of 0.698 mm was obtained. The stiffened plates mesh had the same characteristics as the reference plate's mesh. The central deflection of stiffened plates are shown in Figure 7 as a function of the degree of freedom $h_s/t_s$ for each geometry studied in this work, following the notation $N_{ls} \times N_{ts}$.

In Figure 7, one can notice that there is a pattern in the behavior of the stiffened plates, which enabled the central deflection of the plates to be estimated through equations obtained from a nonlinear regression. The coefficient of determination ($R^2$) of this regression, which is the statistical measure of how well the regression curve matches with the real data, was higher than 0.99 for all combinations of

$N_{ts}$ and $N_{ls}$, indicating an excellent fit. The central deflections (in mm) can be obtained through the following equation:

$$w = \alpha \left(\frac{h_s}{t_s}\right)^\beta \quad (2)$$

where $\alpha$ and $\beta$ coefficients are dependent on the degrees of freedom, i.e., the number of longitudinal and transverse stiffeners. These coefficients and $R^2$ are presented in Table 1 for each combination of $N_{ts}$ and $N_{ls}$ analyzed.

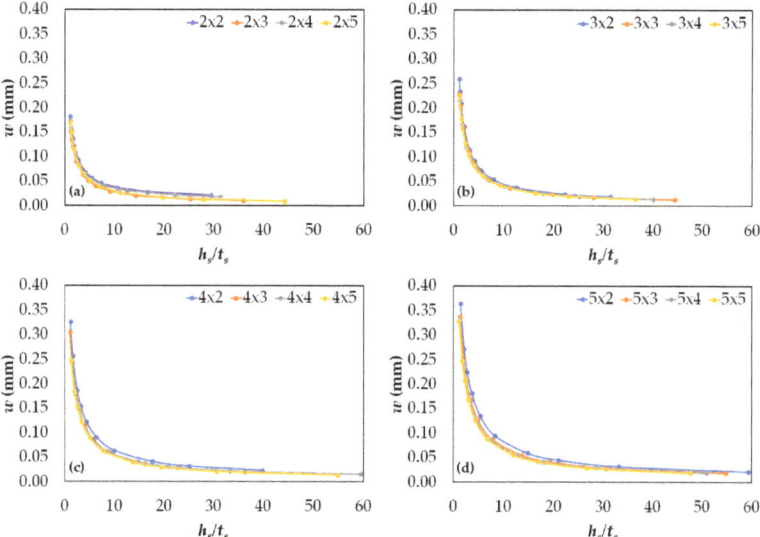

**Figure 7.** Central deflections of the stiffened plates due to $h_s/t_s$ variation with (a) $N_{ls} = 2$ and $N_{ts} = 2$ to 5, (b) $N_{ls} = 3$ and $N_{ts} = 2$ to 5, (c) $N_{ls} = 4$ and $N_{ts} = 2$ to 5, and (d) $N_{ls} = 5$ and $N_{ts} = 2$ to 5.

**Table 1.** Coefficients $\alpha$, $\beta$, and $R^2$ for each combination of $N_{ls}$ and $N_{ts}$.

| $N_{ls} \times N_{ts}$ | $\alpha$ | $\beta$ | $R^2$ |
|---|---|---|---|
| 2 × 2 | 0.1867 | −0.6664 | 0.9961 |
| 2 × 3 | 0.1777 | −0.7881 | 0.9997 |
| 2 × 4 | 0.1782 | −0.6758 | 0.9967 |
| 2 × 5 | 0.1855 | −0.7855 | 0.9997 |
| 3 × 2 | 0.2789 | −0.7754 | 0.9996 |
| 3 × 3 | 0.2527 | −0.7782 | 0.9995 |
| 3 × 4 | 0.2491 | −0.7754 | 0.9994 |
| 3 × 5 | 0.2484 | −0.7842 | 0.9984 |
| 4 × 2 | 0.3755 | −0.7549 | 0.9990 |
| 4 × 3 | 0.3369 | −0.7779 | 0.9974 |
| 4 × 4 | 0.3169 | −0.7461 | 0.9994 |
| 4 × 5 | 0.3153 | −0.7766 | 0.9986 |
| 5 × 2 | 0.4862 | −0.7663 | 0.9983 |
| 5 × 3 | 0.4281 | −0.7723 | 0.9976 |
| 5 × 4 | 0.4004 | −0.7674 | 0.9971 |
| 5 × 5 | 0.3833 | −0.7707 | 0.9959 |

From an analysis of the data presented in Figure 7 and Table 1, it can be noted that an increment in the ratio $h_s/t_s$ caused a decrease in the central deflection for the stiffened plates. This mechanical behavior can be explained by the fact that the moment of inertia increases due to an increase in

$h_s/t_s$, and consequently so does the plate's stiffness. However, for values of $h_s/t_s \geq 20$, there was no significant reduction in the plate's central deflection.

Besides, it is possible to observe that the degree of freedom $N_{ls}$ has more influence on the structural mechanical behavior than $N_{ts}$, as the stiffened plates with the same values of $N_{ls}$ presented similar central displacement when $h_s/t_s$ varied, independently of $N_{ts}$. This fact is evidenced by the similarity of the coefficients $\alpha$ and $\beta$ for plates with the same number of longitudinal stiffeners (see Table 1). Therefore, it is possible to neglect the influence of $N_{ts}$ on the analysis, enabling an estimation of the central deflection of the plates for each $N_{ls}$ through the equations obtained from a nonlinear regression, as a function of exclusively the ratio between the height and the thickness of the stiffeners. Therefore, the central out-of-plane displacements of the stiffened plates with $N_{ls}$ = 2, 3, 4, and 5 can be obtained, respectively, through the following equations:

$$w_2 = 0.181 \left(\frac{h_s}{t_s}\right)^{-0.735} \qquad (3)$$

$$w_3 = 0.256 \left(\frac{h_s}{t_s}\right)^{-0.776} \qquad (4)$$

$$w_4 = 0.332 \left(\frac{h_s}{t_s}\right)^{-0.761} \qquad (5)$$

$$w_5 = 0.412 \left(\frac{h_s}{t_s}\right)^{-0.759} \qquad (6)$$

where $w_2$, $w_3$, $w_4$, and $w_5$ are the central deflections for the plates with $N_{ls}$ = 2, 3, 4, and 5, respectively.

Equations (3)–(6) are even more useful for small values of $h_s/t_s$ where the coefficient that precedes them becomes more relevant. Thus, according to the observed trend, when it was kept constant the total material volume and the ratio $h_s/t_s$ assumed small values, a shorter number of stiffeners led to a smaller deflection.

From Equations (3)–(6) it is also possible to propose a unique equation to represent these equations, given by:

$$w_{N_{ls}} = 0.085 N_{ls} \left(\frac{h_s}{t_s}\right)^{-0.758} \qquad (7)$$

where coefficient $\beta = -0.758$ is the average value for the $\beta$ coefficients of Equations (3) to (6) with a standard deviation of 0.017; the $\alpha$ coefficient, which was determined as the average value of the ratios $\alpha/N_{ls}$ for Equations (3) to (6), is 0.085 (with a standard deviation of 0.004), which should be multiplied by the number of longitudinal $N_{ls}$ stiffeners in each case.

Therefore, to prove the effectiveness of the proposed equations, the numerical results generated in the present work (see Figure 7) were compared with the analytical results from Equations (3) to (6) and Equation (7), being shown in Figure 8.

Equations (3)–(6) had $R^2$ coefficients of 0.9786, 0.9965, 0.9913, and 0.9895, respectively, in relation to the results of the performed numerical simulations, as can be viewed in Figure 9.

In Figure 9, one can note the good agreement of Equation (7) (dashed red curve) with Equations (3)–(6) (continuous green curve). Accordingly, if only Equation (7) be adopted to predict the central deflections of the analyzed stiffened plates a good agreement is achieved, being this simplification more attractive for practical design purposes.

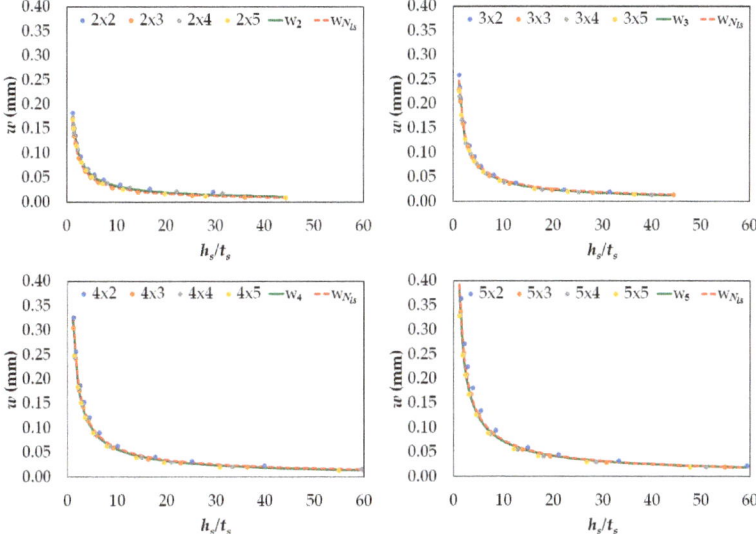

**Figure 8.** Numerical results versus analytical results for the central deflection of stiffened plates.

It is also important to highlight that when the deflections of the stiffened plates are compared to the reference plate, all of the suggested geometrical configurations of the stiffened plates presented a smaller central deflection than that obtained by the non-stiffened plate with the same steel volume. Even though these are all better results, there is great variation in their values as the parameters changed, varying the reduction of the central deflection from 47.68% to 98.57%. This shows how important this type of analysis is. Also, an increment in the number of stiffeners is not directly related to an increase in the structural rigidity because, in order to keep the same material volume of the structure, a greater number of stiffeners entails a reduction in their height and consequently in the moment of inertia.

Lastly, through a global comparison among all analyzed stiffened plates (see Figure 7), it was possible to determine the optimal geometric configuration that would lead to the best mechanical behavior, i.e., the geometric configuration that minimizes the central deflection. The minimal central displacement among all studied geometries was 0.00997 mm and was obtained by the plate with $N_{ls} = 2$, $N_{ts} = 5$, and $h_s/t_s = 44.40$, which is depicted in Figure 9a. The optimal stiffened plate's geometric configuration enabled a reduction of 98.57% in the central deflection if compared to the reference plate. Figure 9 also shows a comparison between the stiffened plates that led to the optimal (Figure 9a) and worst (Figure 9b) geometrical configurations in terms of central deflection. One can observe that in the worst configuration, the stiffened plate had a deformed pattern similar to a plate without stiffeners, i.e., the maximal deflection occurred in the plate's central point; for the optimized geometry, the stiffeners, mainly due to its $h_s/t_s$ value, promoted a division of the plate into several regions, each behaving like an "unstiffened" small plate. Moreover, it is evident that the magnitude of the deflections was heavily affected by the variation of $h_s/t_s$, $N_{ls}$, and $N_{ts}$ once the same material quantity was adopted in all cases.

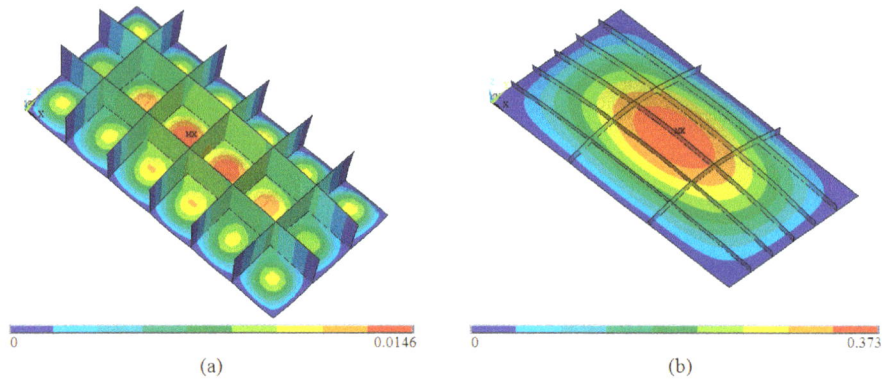

**Figure 9.** Deformed configuration of (**a**) the optimal plate and (**b**) the worst plate (in mm).

## 5. Conclusions

In the present work, a structural numerical analysis was performed to estimate the central deflection of thin stiffened plates and the influence of parameters such as the number of longitudinal and transverse stiffeners and the ratio between their height and thickness. To do so, it was kept constant the total volume of the plate.

From the results obtained, as expected, the application of stiffeners in thin plates provided greater rigidity to the structure. All stiffened plates presented central displacements that were smaller than the one presented by the non-stiffened plate used as a reference. Moreover, the importance of performing studies on this subject was shown, since the geometric configuration of the stiffeners had a substantial influence on the deflection values. Wide variations were shown in these values due to modifications in the geometry. It was also noted that for $h_s/t_s$ values greater than 20, there was no significant reduction in the central deflection of the stiffened plates analyzed.

Furthermore, from the results of this work, it was possible to obtain, through nonlinear regressions, equations that accurately describe the deflection of the studied stiffened plates. From these equations, along with an analysis of the charts, it was observed that the out-of-plane central displacement decreased as the ratio $h_s/t_s$ increased. In addition, by using these equations, the deflection values could be estimated based exclusively on the ratio between the stiffeners' heights and thicknesses ($h_s/t_s$) for each $N_{ls}$ value. These deflection values were even more useful when this relation assumed small values. In addition, a simple and effective equation was proposed to determine the central deflections of the stiffened plates studied in the present work.

Finally, the global optimized geometric configuration was determined among all analyzed plates, i.e., the stiffened plate geometry that minimized the central deflection. Thus, the optimized geometry was the one that presented two longitudinal stiffeners, five transverse stiffeners, and a ratio $h_s/t_s$ of 44.40. This geometry enabled a reduction of 98.57% in the central deflection when compared with the non-stiffened reference plate.

In future work, it is recommended to investigate other values of ∅, as well as other types of stiffeners. It would also be interesting to perform a complementary study to this work where not only the central deflection but also the mechanical behavior regarding the stresses be taken into account.

**Author Contributions:** Data curation, J.P.T.P.d.Q.; Formal analysis, L.A.I.; Investigation, J.P.T.P.d.Q., M.L.C., and G.d.S.T.; Methodology, L.A.O.R., E.D.d.S., and L.A.I.; Resources, A.P.; Supervision, L.A.O.R., E.D.d.S., and L.A.I.; Validation, G.d.S.T.; Writing—original draft, J.P.T.P.d.Q.; Writing—review & editing, M.L.C., A.P., E.D.d.S., G.d.S.T., and L.A.I.

**Funding:** This research was funded by FAPERGS (Research Support Foundation of Rio Grande do Sul, *Porto Alegre*, RS, Brazil) grant number [Edital PROBITI 2018-2019] and by CNPq (National Council for Scientific and Technological Development—Brasília, DF, Brazil) grant number [306012/2017-0].

**Acknowledgments:** The authors acknowledge the Brazilian agencies FAPERGS (Research Support Foundation of Rio Grande do Sul) and CNPq (National Council for Scientific and Technological Development—Brasília, DF, Brazil) for the financial support. The authors E.D. Santos, L.A. Isoldi, and L.A.O. Rocha are grant holders of CNPq.

**Conflicts of Interest:** The authors declare no conflict of interest.

## References

1. Mansour, A.; Liu, D. *The Principles of Naval Architecture Series: Strength of Ships and Ocean Structures*, 1st ed.; The Society of Naval Architects and Marine Engineers: Jersey City, NJ, USA, 2008; p. 251, ISBN 0-939773-66-X.
2. Karimirad, M.; Michailides, C.; Nematbakhsh, A. *Offshore Mechanics: Structural and Fluid Dynamics for Recent Applications*, 1st ed.; Wiley: Hoboken, NJ, USA, 2018; p. 297, ISBN 9781119216643.
3. Rossow, M.P.; Ibrahimkhail, A.K. Constraint method analysis of stiffened plates. *Comput. Struct.* **1978**, *8*, 51–60. [CrossRef]
4. Bedair, O.K. Analysis of stiffened plates under lateral loading using sequential quadratic programming (SQP). *Comput. Struct.* **1997**, *62*, 63–80. [CrossRef]
5. Sapountzakis, E.J.; Katsikadelis, J.T. Analysis of plates reinforced with beams. *Comput. Mech.* **2000**, *26*, 66–74. [CrossRef]
6. Silva, H.B.S. Análise Numérica da Influência da Excentricidade na Ligação Placa-viga em Pavimentos Usuais de Edifícios. Master's Thesis, Universidade de São Paulo, São Carlos, Brazil, 2010.
7. Troina, G.S.; de Queiroz, J.P.T.P.; Cunha, M.L.; Rocha, L.A.O.; dos Santos, E.D.; Isoldi, L.A. Verificação de modelos computacionais para placas com enrijecedores submetidas a carragamento transversal uniforme. *Revista CEREUS* **2018**, *10*, 285–298. [CrossRef]
8. Bejan, A.; Lorente, S. *Design with Constructal Theory*; John Wiley & Sons: Hoboken, NJ, USA, 2008.
9. Rodrigues, M.K.; Brum, R.S.; Vaz, J.; Rocha, L.A.O.; dos Santos, E.D.; Isoldi, L.A. Numerical investigation about the improvement of the thermal potential of an earth-air heat exchager (EAHE) employing the constructal design method. *Renew. Energy* **2015**, *80*, 538–551. [CrossRef]
10. Helbig, D.; da Silva, C.C.C.; Real, M.V.; dos Santos, E.D.; Isoldi, L.A.; Rocha, L.A.O. Study about buckling phenomenon in perforated thin steel plates employing computational modeling and Constructal design. *Latin Am. J. Solids Struct.* **2016**, *13*, 1912–1936. [CrossRef]
11. Szilard, R. *Theories and Applications of Plate Analysis: Classical, Numerical and Engineering Methods*; John Wiley & Sons: Hoboken, NJ, USA, 2004.
12. Song, X.; Dai, J.X. Mechanical Modeling and ANSYS Simulation Analysis of Horizontally Axial Wind Turbine Tower. *J. Gansu Sci.* **2011**, *1*, 028.
13. Zainal, S.; Tan, C.S.C.; Siang, T. ANSYS simulation for Ag/HEG hybrid nanofluid in turbulent circular pipe. *J. Adv. Res. Appl. Mech.* **2016**, *23*, 20–35.
14. Pavlovic, A.; de Camargo, F.V.; Fragassa, C. Crash safety design: Basic principles of impact numerical simulations for composite materials. *AIP Conf. Proc.* **2018**, *1981*, 020032. [CrossRef]
15. De Camargo, F.V.; Pavlovic, A. Fracture evaluation of the falling weight impact behaviour of a basalt/vinylester composite plate through a multiphase finite element model. *Key Eng. Mater.* **2017**, *754*, 59–62. [CrossRef]
16. Boria, S.; Pavlovic, A.; Fragassa, C.; Santulli, C. Modeling of falling weight impact behavior of hybrid basalt/flax vinylester composites. *Procedia Eng.* **2016**, *167*, 223–230. [CrossRef]
17. Pavlovic, A.; Fragassa, C.; Disic, A. Comparative numerical and experimental study of projectile impact on reinforced concrete. *Compos. Part B* **2017**, *108*, 122–130. [CrossRef]
18. Fragassa, C.; de Camargo, F.V.; Pavlovic, A.; Silveira, A.C.F.; Minak, G.; Bergmann, C.P. Mechanical characterization of gres porcelain and low-velocity impact numerical modeling. *Materials* **2018**, *11*, 1082. [CrossRef] [PubMed]
19. Da Silva, P.C.; Ramos, A.P.; Lima, J.P.S.; Junior, M.C.B.P.; Rocha, L.A.O.; dos Santos, E.D.; Real, M.V.; Isoldi, L.A. Simulação numérica e constructal design aplicados à melhoria do comportamento mecânico de placas finas de aço com enrijecedores submetidas à flexão. In Proceedings of the CILAMCE, Rio de Janeiro, Brazil, 22–25 November 2015.

20. Lima, J.P.S.; Rocha, L.A.O.; dos Santos, E.D.; Real, M.V.; Isoldi, L.A. Constructal design and numerical modeling applied to stiffened steel plates submitted to elasto-plastic buckling. *Proc. Romanian Acad. Ser. A* **2018**, *19*, 195–200.
21. ANSYS. *ANSYS Mechanical APDL Basic Analysis Guide*; ANSYS, Inc.: Canonsburg, PA, USA, 2008.
22. Timoshenko, S.; Woinowsky-Krieger, S. *Theory of Plates and Shell*, 2nd ed.; McGraw-Hill: New York, NY, USA, 1964.

 © 2019 by the authors. Licensee MDPI, Basel, Switzerland. This article is an open access article distributed under the terms and conditions of the Creative Commons Attribution (CC BY) license (http://creativecommons.org/licenses/by/4.0/).

*Review*

# Survey on Experimental and Numerical Approaches to Model Underwater Explosions

**Felipe Vannucchi de Camargo**

University of Bologna, Interdepartmental Center for Industrial Research on Advanced Mechanics and Materials, Viale del Risorgimento 2, 40136 Bologna, Italy; felipe.vannucchi@unibo.it; Tel.: +39-051-209-3266

Received: 29 October 2018; Accepted: 8 January 2019; Published: 15 January 2019

**Abstract:** The ability of predicting material failure is essential for adequate structural dimensioning in every mechanical design. For ships, and particularly for military vessels, the challenge of optimizing the toughness-to-weight ratio at the highest possible value is essential to provide agile structures that can safely withstand external forces. Exploring the case of underwater explosions, the present paper summarizes some of the fundamental mathematical relations for foreseeing the behavior of naval panels to such solicitation. A broad state-of-the-art survey links the mechanical stress-strain response of materials and the influence of local reinforcements in flexural and lateral-torsional buckling to the hydrodynamic relations that govern the propagation of pressure waves prevenient from blasts. Numerical simulation approaches used in computational modeling of underwater explosions are reviewed, focusing on Eulerian and Lagrangian fluid descriptions, Johnson-Cook and Gurson constitutive materials for naval panels, and the solving methods FEM (Finite Element Method), FVM (Finite Volume Method), BEM (Boundary Element Method), and SPH (Smooth Particle Hydrodynamics). The confrontation of experimental tests for evaluating different hull materials and constructions with formulae and virtual reproduction practices allow a wide perception of the subject from different yet interrelated points of view.

**Keywords:** stiffened plate; constitutive model; finite element; fluid-structure interaction; ship design; state-of-the-art

## 1. Introduction

Acknowledging how to properly soften the effects of impact-related damage is an imperative design guideline in shipbuilding. Specially for military applications, underwater explosions (UNDEX) prevenient from subsea blast loads can infer irreparable structural impairment to vessels, where the dynamic response of ships depends on the influence of several parameters, such explosive power, blast distance, hull panel composition, and reinforcements. The mechanics involved in ship collisions and impact can thus be looked at from two inter-dependent perspectives [1], where the external mechanics describe the hydrodynamics surrounding the vessel and the magnitude with which they affect it, and the internal mechanics that regard how the materials respond to these forces and dissipate energy through strain.

The necessary comprehension of the materials behavior goes beyond traditional steels, whereas modern designs utilize advanced metallic alloys [2], composite structures (either purely polymeric, as in the case of small and medium sized boats [3], or in a sandwich layout composed by metal sheets and foam cores [4]) and wood [5]. Also, the usage of innovative materials, such as polymers reinforced by natural fibers, have been addressed in the literature, on one hand presenting inferior mechanical properties to carbon or glass fiber, and on the other having enhanced sustainability [6]. In addition, the diverse existent hull architecture requires the designer not only to master different materials, but also to understand the response of the vessel itself as a composition of the several interlinked structural elements built with those materials, such as panels, stiffeners, and girders. The way in which these structures are realized and arranged ultimately determines the resistance of the ship and define if reinforcements like stiffeners and girders apply, and what is the optimal geometry for plate cutouts (e.g., for hatches) to avoid undesired elastoplastic deformations to buckling [7–9].

External mechanics, in turn, involves a series of differential hydrodynamic relations that describe the propagation of intermittent pressure waves caused by explosions defining the velocity, temperature, period, and force with which hulls are hit. The consequent fluid-structure interactions with relevant severity do not only involve liquid matter; gas bubbles generated by the blasts can also cause noticeable damage given the high pressure and temperature they may present [10,11].

The relevance of UNDEX for military purposes is clear, as highlighted in studies dated from 1990 [12], which explicitly express the cold war as the impetus for such research initiative, stating as border condition for the case study a soviet missile impacting a double-hulled structure. The aforementioned work defined through a series of experiments the strongest parameters that influence UNDEX effects as being: enthalpy of detonation, number of moles of gas, molecular weight of gas, and solid phase density.

The strategic usage of the arctic ocean as an operational area for submarines has also had its importance reflected by studies such as Barash's [13], which analyze the effect of UNDEX beneath artic ice, not only to improve the safety of those vessels to enemy blasts, but also to allow them to have quick access to the surface by breaking ice packs using explosives. Potentially harmful debris originated from the blasts were addressed by Bryant [14], concluding that little or no metallic debris are deposited in the medium during early bubble oscillations, whereas they are actually transported by the bubbles and released in a certain area as the oscillations terminate.

Early analyses on the physics of the pressure waves were also performed. Snay [15] confirmed that the pulse shape of waves varies during their propagation in the water, where the steepness of the pressure at the head of the wave rises until the formation of an impact front. However, the author states the validity of approximating UNDEX to acoustic waves (which do not change their shape) for the solution of interaction problems, modelling them as small-amplitude constant-sound-velocity waves.

The scientific way of approaching UNDEX problems itself is an issue with distinct propositions in precursor military-grade reports. O'Daniel et al. [16] evaluated the mechanics of bubble jets, suggesting that such analysis could be conducted by performing three experiments of bubble jet strike on vertical targets in different scales: a small-scale one to study the phenomenology of the bubble jet; a mid-scale one to measure the loads applied by these jets on targets; and a large-scale one for the structural response of targets. Miller [17] categorized three fundamental steps: comparison of structural response to conventional and nuclear pressure profiles, followed by finite element analysis, and then by writing an appropriate computational numerical code to describe the experiments. Naturally, since then, computer-aided simulations have been immensely improved, as shown hereby.

Built upon classic mathematical models, numerical simulation software is a well-established time-saving tool for yielding precise UNDEX results, leaving aside the need for resource costly experimental tests. In marine engineering, accurate virtual reproductions can be done for UNDEX provoked pressure waves [18] and bubbles [19] over vessels, as well as water blast on sandwich panels [4] and cavitation in propellers [20]. For that, a variety of constitutive material models and numerical solving techniques may be used. Given the importance of understanding how external and internal mechanics work and relate, the present work briefly summarizes the fundamental numerical expressions of each one, giving also a glance in how they are usually represented in numerical simulations.

Portraying the three independent but correlated parts of UNDEX described by the navy as shown in Brenner et al. [21], i.e., the responses of water, of the vessel structure, and of the materials to the shock waves, Figure 1 depicts through a flow chart scheme the aspects approached in this work in an orderly manner. In other words, the current state-of-the-art explains how the pressure impulse propagates through water and with which magnitude it hits the ship (taking into account pressure waves and bubbles), how the ship structure responds to this impulse, and in which way the material components of the hull absorb this energy, deform, and eventually fail.

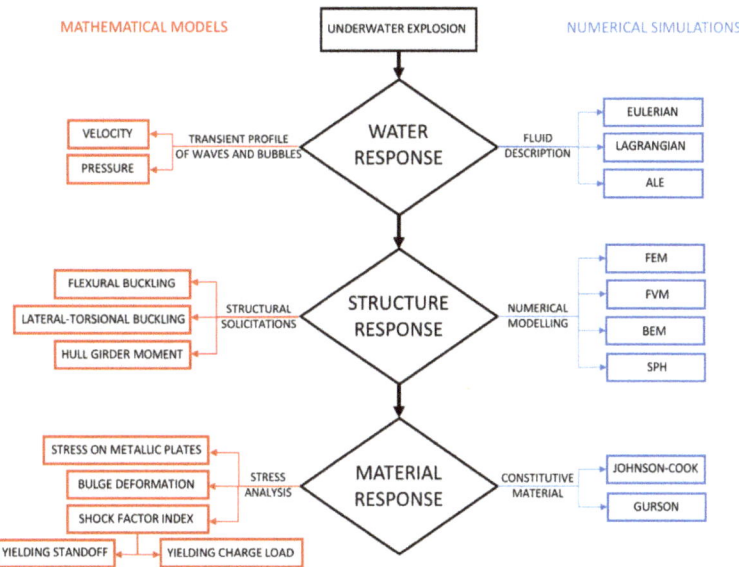

**Figure 1.** Flow chart scheme of UNDEX aspects approached in the present work.

The characterization of such phenomena is hence made through analytical models, and directions on how to reproduce all the three steps by numerical simulation are also given. Some fundamental notions are provided on Eulerian, Lagrangian, and Aleatory Lagrangian-Eulerian (ALE) fluid descriptions on Finite Element and Volume Methods (FEM and FVM, respectively), Boundary Element Method (BEM), and Smoothed Particle Hydrodynamics (SPH) numerical modelling techniques, and on Johnson-Cook's and Gurson's constitutive materials.

Acknowledging that the study of UNDEX phenomena is an intricate and wide field of study with deeper aspects to be considered, some of its basic yet most important parameters are hereby summarized and briefly explained, intending to allow the comprehension of fundamental notions of the subject based on a thorough literature survey.

## 2. Analytical Models

### 2.1. Hydrodynamics and Fluid-Structure Impact

Amongst the challenges in ship design, there are several forces to be taken into account that may act either separately or combined over the structure originated from gravity, wind pressure, friction, freight, and hydrostatic lifting force, which can infer massive shear and torsional efforts given the large size of a ship and due to its weight distribution. Moreover, fluid-structure interactions (FSI), such as pressure waves, gas bubbles, cavitation, and underwater explosions also have the potential to cause significant damage, given that water transmits explosive energy much more efficiently than air [16].

Differently from solid-particle impacts, where a specific damage is locally caused by a particularly shaped projectile [22], in UNDEX, all the extension of the hull reachable by the same pressure wave is approximately equally impaired. These spherical waves originate a mass flow behind them (i.e., the afterflow), which leads to the formation of bubbles that can be regular UNDEX bubbles, steam bubbles in the case of nuclear explosions, or cavitation bubbles in the case of explosions that do not provoke gases [15].

Bubbles generated by UNDEX, for instance, which constitute hot air masses that can hit a vessel in more than 10 consecutive waves [23], can reach impact pressures of up to 15% of the wave pressure [11], causing significant damage and even a potential and dangerous match in resonance frequencies with the ship. Furthermore, Cui et al. [24] experimentally analyzed the mechanics of these bubbles, finding out that the pressure peaks they induce on the impacted body is highly dependent on their shape before collapse, given that the less spherical and more asymmetrical it is, the smaller is the pressure due to the influence of splashing jet water. The shape of bubbles, as shown by Gong et al. [19], is directly related to the position of the explosive charge in relation to the hull.

The response to underwater explosions is a topic of interest for military purposes, given that even if the hull is not directly hit by a torpedo or collided to other structures, the sole propagation of energy from a far blast center through water is capable of causing significant damage to the ship. This energy release can include gas bubbles at nearly 3000 °C and pressures up to 5 GPa, besides solid particles made from lead or alumina, for instance, which altogether may impact the hull's surface at nearly the velocity of sound [10].

It is possible to quantify the pressure in the proximity of the wave front $P(t)$ if the charge is detonated at less than 100 m in depth [25] by the expression of Equation (1) [10]:

$$P(t) = P_m \begin{cases} e^{-t/\theta} & if \quad t < \theta \\ \frac{0.368\theta}{t} & if \quad \theta < t < t_1 \end{cases} \quad (1)$$

where the time constant $\theta$ can be defined by Equation (2) [10]:

$$\theta = R_0 \begin{cases} 0.45 \, \bar{r} \, 10^{-3} & if \quad \bar{r} \leq 30 \\ \frac{3.5}{c}\sqrt{\log \bar{r} - 0.9} & if \quad \bar{r} > 30 \end{cases} \quad (2)$$

where $\bar{r}$ is the ratio between the distance of the explosion center and the quotient of the measured location $R$ by the initial radius of the spherical explosive $R_0$. The peak pressure $P_m$ (MPa) and the time decay constant $\theta$ (ms) can also be defined as a function of the charge load $W$ and the standoff distance $S$ (Equations (3) and (4), respectively) [26]:

$$P_m = \alpha \left(\frac{W^{1/3}}{S}\right)^\beta \quad (3)$$

$$\theta = 0.058 W^{1/3} \left(\frac{W^{1/3}}{S}\right)^{-0.22} \quad (4)$$

where $\alpha$ and $\beta$ are constants, which highlights that it is important to notice the discrepancies found in the literature for them. While [26,27] adopt $\beta$ as 1.13, [18] considers 1.81. By its turn, $\alpha$ is adopted as 148.93 in [26], 52.16 in [27], and 29.9 in [18]. In addition, [26] considers Equation 3 valid when the measuring point is located until 10 times the explosion radius, while [18] states that the adequate tolerance distance is up to 6 times the charge radius.

When the underwater wave passes through a certain point in space, it is then submitted to a transient pressure $P(t)$ and displaced in a velocity $v(t)$ in the direction of the flow. Considering a spherical flow, it is possible to estimate and correlate both into the same time-dependent Equation (5) [26], where the first term stands for the velocity of a plane wave, and the second, called "after flow" term, is important for large time intervals and for when the measuring point is close to the explosions, i.e., the standoff is low.

$$v(t) = \frac{P(t)}{\rho c} + \frac{1}{\rho S}\int_0^t p(t)dt \quad (5)$$

Although the strongest shock wave comes along with the first UNDEX impact, it is important to acknowledge that the explosion damage is characterized by repetitive impacts originated in the explosion center, which intensity decays exponentially when hitting a marine structure [28], and the second impact can even be the most damaging one due to the bubble pulse, especially to air-backed hull panels [29]. This concept is illustrated in Figure 2.

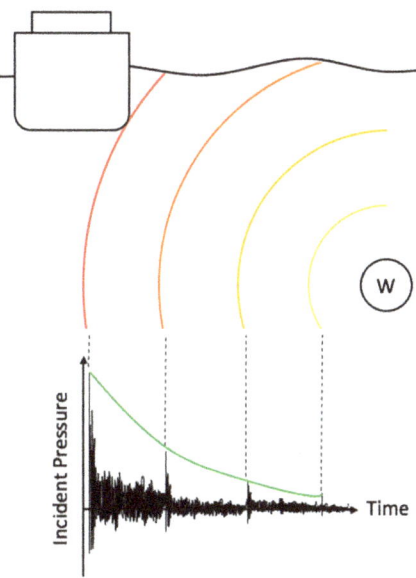

**Figure 2.** Successive underwater wave impacts from the explosion load (W) and incident pressure decay.

## 2.2. Structural Response

Structural elements are meant to be strategically assembled in the construction of ships in places where they can aggregate normal, shear, or torsional resistance to the vessel, preferably in profiles and shapes that favor their mechanical properties and account for the lowest weight attainable. The main structures are longitudinal and transversal girders and stiffeners, providing enhanced resistance to underwater or contact explosions compared to unstiffened vessels [30]. These can grant noticeable economy without compromising strength and durability if stiffeners are adequately positioned at regular gaps, while dividing the hull shell into lattices in which the intersection point between transversal and longitudinal girders are the most resistant zones [31].

Hung et al. [30] have demonstrated experimentally the importance of stiffeners by performing underwater explosion tests on 3 cylindrical specimens, one being unstiffened, one externally, and one internally stiffened with welded rings. The latter showed much smaller strain, pressure, and fluid acceleration inside the coupon. Ming et al. [18] have complemented this approach by numerically and experimentally studying the reaction of plates reinforced with "T" profile stiffeners placed in the opposite side of an underwater explosion, demonstrating the damage steps of the reinforced plate. The study states that after both the plate and the stiffeners are initially damaged due to pressure waves, the panel in the lattice is torn, leading to the collapse of the longitudinal and transversal stiffeners joints, weakening the whole structure significantly.

Gordo et al. [32] have developed a numerical method to estimate the ultimate longitudinal strength of the hull girder, yielding accurate confrontations against similar approaches and the real ship failure. The compression strength of stiffened plate columns at a determined strain is defined by two independent expressions that describe flexural buckling ($\phi_F$) and lateral-torsional buckling (or tripping) ($\phi_T$). The first is described by Equation (6):

$$\phi_F(\varepsilon_i) = \phi_{JO}(\varepsilon_i)\frac{A_s + w_e bl}{A_s + bl} \qquad (6)$$

where $\phi_{JO}$ is based on the Johnson-Ostenfeld formulation that considers inelastic effects during buckling [33], and the second term of the equation stands for the actual degradation of the plate due to compression loading. $A_s$ is the sectional area, $b$ is the breadth of plating between longitudinal stiffeners, $l$ is the length of the stiffeners, and $w_e$ is the effective width of the plate given by Equation (7):

$$w_e = max(-1, min[1, \varepsilon_i])\left(\frac{2}{\beta} - \frac{1}{\beta^2}\right) \qquad (7)$$

where $\beta$ is the slenderness of the plate equal to $\beta = (b/l).\varepsilon_v$, in which the second term represents the instantaneous strain. In turn, the tripping strength can be calculated by Equation (8):

$$\phi_T(\varepsilon_i) = \phi_{Tmin}\frac{\varepsilon_t}{\varepsilon_i}\frac{A_s + w_e bl}{A_s + bl} \qquad (8)$$

The first term stands for the maximum elastic tripping stress, and the second to the strain of maximum load. It is important to highlight that the resistance of the girder can be affected by residual stress and external agents such as corrosion [32], and that although both influence the behavior of the material, they should not be taken into account simultaneously, given that while residual stress acts mainly in the beginning of the operational life of the ship, corrosion is negligible in the beginning and increases with time. To avoid lateral-torsional bending in ships, the influence of the residual stress over the ultimate bending moment is quantified in Equation (9):

$$M_{ur} = (1 - 0.3\bar{\sigma}_r)M_u \qquad (9)$$

where $\bar{\sigma}_r$ is the residual stress of the panels in compression and $M_{ur}$ and $M_u$ are, respectively, the ultimate bending moment with and without residual stress.

Innovative mechanical reinforcement techniques that go beyond classic metallic structures have also been studied, such as rubber coated ships [34], which show the ability of damping the high-frequency response from pressure waves, helping to keep the integrity of the vessel to crashes and close-range underwater explosions that may cause elastic and even plastic deformations to the side structures. However, this coating is not so effective regarding low-frequency whipping motions caused by gas bubbles.

### 2.3. Material Response

The first important material feature that comes to mind when one needs a vessel design to withstand impact and wear solicitations is high mechanical resistance thresholds [35,36]. However, in addition to being tough, building materials must also be as lightweight as possible to consume less fuel, leave room for more load bearing capacity and to provide the ability to quickly overcome inertia when maneuvering, especially in the case of military vessels, to make them difficult targets.

The study of metallic phases and alloys figure as an important resource to understand and improve the resistance of military naval structures to underwater explosions [37], especially in terms of the high inherent density attributed to metals. Latourte et al. [2] compared experimentally and numerically high-strength martensitic and austenitic alloys, investigating their deformation and fracture characteristics, where the first was designed to present higher fracture toughness and maximum strength, while the latter has a better uniform ductility, avoiding premature necking and consequent localized failure. The interesting outcome is that the martensitic steel was found out to sustain underwater fluid-structure impacts better, where the force magnitude needed to infer failure is 25% higher than for austenitic steel (although the austenitic performance can be improved by tempering) [2].

Besides traditional monolithic metals, the coupling of different materials in sandwich layouts have presented interesting features relevant for ship design, like the flexibility in the core constitution, which can lead to a stiffer overall material in the through-the-thickness direction, or can soften and dampen impact and decrease the force transmission to the supporting structure [4]. Furthermore, sandwich panels present an enhanced performance against underwater fluid shocks when compared to single-constituent elements of equal weight [38,39].

Fan et al. [40] investigated the response of sandwich panels with aluminum sheet skin and a honeycomb core to underwater blasts, comparing the results with monolithic plates of equivalent mass, proving a better performance of the sandwich both in terms of deformation resistance and secondary pressure wave intensity. The first advantage can be furthermore improved by increasing the equivalent thickness of the composite. As for composites with polymeric matrix, Gong et al. [19] studied the transient behavior of glass-epoxy composites to UNDEX, showing that although a composite hull might amortize the effects of bubble impacts better than steel, it becomes more susceptible to global mechanical effects, not only local ones as would be the case for metals.

In an experimental underwater explosion test, it is also possible to identify the energy absorption capacity of a certain material by analyzing its deflection, once it has been proved that during plastic regime deformation, the ability of the material to absorb energy is proportional to the square of its deflection [23]. Another important aspect is the specimen geometry to be adopted in a bulge test, given that parabolic and spherical shapes absorb practically the same energy, while conical and hyperbolic absorb only half of that considering the same bulge depth [27]. Regarding experimental tests, it is possible to predict the maximum von Mises stress for both thin circular ($\sigma_c$) [41] and rectangular plates ($\sigma_r$) [42], as shown respectively in Equations (10) and (11), considering that it is an explosion with low intensity (for specimens farther from the charge more than 10 times its explosion radius) and that the deformation is realized within the elastic regime.

$$\sigma_c = \sqrt{\frac{6E\rho_P P_m^2 \chi^{2/(1-\chi)}}{\rho^2 c^2 (1-v)}} \tag{10}$$

$$\sigma_r = 0.867 \sqrt{\frac{14 E \rho_P P_m^2 \chi^{2/(1-\chi)}}{\rho^2 c^2}} \tag{11}$$

where E is the elastic modulus and $\rho_P$ is the density of the plate material, $\rho$ is the density of water, c is the velocity of sound in the water medium, $P_m$ is the peak pressure, $\chi$ is the angle of incidence to the plate, and $v$ is the Poisson's ratio. In order to indicate the level of damage caused by a shock, the shock factor index (SF) is used as displayed in Equation (12) [42]. To verify whether the shock energy can cause yielding to a thin air backed plate, the yield shock factor ($SF_y$) would have to reach the limiting value expressed by Equation (13) [42].

$$SF = 0.445 \frac{\sqrt{W}}{S} \tag{12}$$

$$SF_y = Y \frac{1}{\sqrt{\eta}} \sigma_y \sqrt{t}, \tag{13}$$

where W is the TNT equivalent of charge quantity (kg), S is the standoff from the explosion center, t is the thickness of the plate, $\sigma_y$ is the yield stress of the plate material, Y is the yield factor equivalent to $2.212 \times 10^{-9}$ for circular plates ($Y_c$), and $1.997 \times 10^{-9}$ for rectangular ones ($Y_r$), and $\eta$ stands for the coupling factor, which is a function of the incidence angle as shown in Equation (14).

$$\eta = 4\chi^{(1+\chi)/(1-\chi)} \tag{14}$$

It is also possible to determine the yielding shock factor ($SF_y$) as a function of the necessary charge quantity $W$ (Equation (15)) or the standoff $S$ (Equation (16)) and a time constant $\theta$ (ms).

$$W = \left(\frac{SF}{0.445}\right)^{0.5946} \left(\frac{\theta}{96.6 \times 10^{-6}}\right)^{2.7026} \quad (15)$$

$$S = \left(\frac{0.445}{SF}\right)^{0.7027} \left(\frac{\theta}{96.6 \times 10^{-6}}\right)^{1.3513} \quad (16)$$

As for the outcome material damage, it mainly depends on the material, boundary conditions, and loading rate considered. Ming et al. [18] have shown through experiments the particular case of clamped flat metal plates, in which damage can be visually categorized in three subsequent steps. The formation of a localized protuberance in the point of maximum stress immediately before the material fracture takes place ("bulging"), leads to detachment of a small metal fragment giving place to a hole ("discing"), whose size is determined by the plate thickness, charge weight, and yield stress [26,43] that continues to grow until the material dissipates this energy through the propagation of radial and equally spaced cracks, deforming the metal in petals ("petaling"), as detailed in Figure 3. A particular aspect noticed by this study is that while the hole size is sensitive to the peak pressure, the deflection near the whole is strongly affected by the impulse of the explosion.

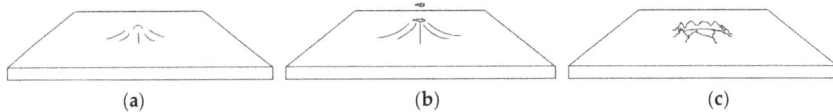

(a)  (b)  (c)

**Figure 3.** Failure steps of bulging (**a**), discing (**b**), and petaling (**c**) for metallic square coupons when subjected to underwater explosion tests.

The hole radius $R$ found in the discing step was calculated by Rajendran et al. [43], as described in Equation (17):

$$R = \sqrt{2\beta W E_i / \pi t \sigma_y \varepsilon_f} \quad (17)$$

where $\beta$ is the ratio of effective work, $E_i$ is the inner energy per unit mass of the explosive, and $\varepsilon_f$ is the fracture strain. Before that, while the material is in the bulging stage, its round-shaped deformation can be calculated by Equation (18) [43], also represented by Figure 4, where $R_S$ and $R_C$ are, respectively, the radii of the sphere and the circle; $t$ is the thickness of the plate; and $D_S$ is the depth of the instantaneous indentation over the plate:

$$R_S = \frac{R_C^2}{2D_S} + \frac{D_S}{2} \quad (18)$$

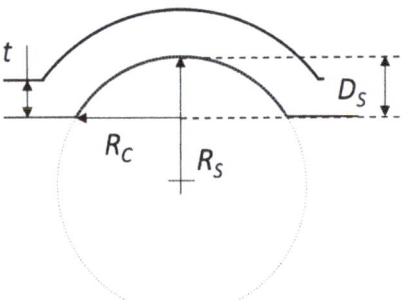

**Figure 4.** Geometrical relation during bulging.

Once the mechanical response of metals to UNDEX is highly dependent on the strain rate of the solicitation, it is vital to illustrate that this effect must be taken into account in calculations for non-idealistic results. Rearranging the stress-strain relations found in [44], this behavior of metallic plates can be summarized in Figure 5:

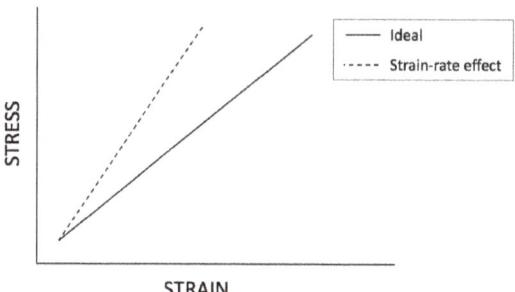

**Figure 5.** Representation of stress-strain behavior of metals when acknowledging the effect of strain rate.

It is possible to see that metallic plates tend to absorb less energy and behave in a more brittle manner when the strain rate effects are acknowledged in stress-strain calculations. This is made by considering a strain rate factor $n$ as a constant associated to stress. The factor $n$ defines the property demeaning level of metallic plates, being directly proportional to the impulse load and inversely proportional to the thickness and yielding stress of the plate. This relation is valid for both circular and rectangular plates [27]. As demonstrated by Jones [44], the influence of $n$ can be calculated by rearranging stress-strain as a relation between the deflection at the center of the plate ($\delta$), the thickness ($t$), and a modified damage parameter ($\phi_M$) that embraces both loading and plate dimensions, through Equations (19) and (20) for circular ($n_c$) and rectangular plates ($n_r$), respectively.

$$\left(\frac{\delta}{t}\right)_c = 0.817 \frac{\phi_M}{\sqrt{n_c}} \tag{19}$$

$$\left(\frac{\delta}{t}\right)_r = 0.95\left[\left(1 + 0.6637\frac{\phi_M^2}{n_r}\right)^{1/2} - 1\right] \tag{20}$$

In those relations, the strain rate factor can be calculated as shown by Equations (21) and (22), where $D$ and $q$ are material parameters, $I$ is the impulse, $t$ is the thickness of the plate, $\sigma_y$ is the yielding stress, $\rho_p$ is the material density, and $R$ is the radius of the circular hole bored.

$$n_c = 1 + \left(\frac{I^2}{3\rho_p^2 t^2 DR}\left(\frac{\rho_p}{3\sigma_y}\right)^{1/2}\right)^{1/q} \tag{21}$$

$$n_r = 1 + 0.0357\left(\frac{I^2}{t^2\sqrt{\sigma_y}}\right)^{1/q} \tag{22}$$

As seen above, small-scale UNDEX experimental tests are routinely carried out by many authors to allow the perception of how materials would behave in operational conditions. However, especially in the case of UNDEX, where high-magnitude forces directly affect materials and structures, involving a complex array of fracture mechanisms, the ability to scale those experiments to the larger and thicker panels that are actually applied in ship building and are subjected to higher stresses is relevant.

However, the intrinsic difficulty to reproduce these experiments in a larger-scale that is closer to reality is clear: high cost, safety concerns, logistics, access to high amounts of explosives, and confidentiality issues make these real-sized tests practically exclusive to national Defense Departments. Therefore, as stated by Cui et al. [24], small-charge experiments remain the best way to study UNDEX phenomena.

Jacob et al. [45] analyzed the scaling effect in regular test scale experiments. Regarding the relation between deflection and impulse, it is reported that the exposed plate area does not influence the deformation, but thickness does. Also, the charge diameter is relevant to determine the plate response. As for the effects of charge height and diameter, the impulse increases as either height or diameter increase independently of the thickness of the plate. This study provides some graphic extrapolation of data to predict the outcome of not-so-practicable tests, but it advises the usage of numerical simulations as a reliable tool to reproduce even operational-scale conditions, meaning that there is no point in performing extensive, dangerous, and costly large-scale experiments.

## 3. Numerical Simulations

Diffused in most engineering fields, numerical simulations represent a very useful tool for faithfully predicting the behavior of materials to various mechanical solicitations, constituting a valid and time-saving approach in designing structures and materials to avoid rework and ensure safety at an optimal cost. Thus, some of the fundamental simulation aspects regarding fluid description, structural, and materials responses to UNDEX will be analyzed.

*3.1. Fluid Description Algorithms*

In numerical simulations, the fluid dynamics can be described by two main approaches, the Eulerian and the Lagrangian. Both can be used to reproduce free fluid surface and deforming wall boundaries, but present some differences. In the first, the properties of the flow are given for defined spatial coordinates as a function of time, by observing how the properties of a certain point in space change due to the flow that passes through it. On the other hand, the Lagrangian method depicts the flow as a large number of individual particles whose motion is described, following them and tracking their property variations over time. These different methods are reproduced within numerical simulations by either setting a fixed grid in space (Eulerian) or nodes that move according to the velocity field in a meshless setup (Lagrangian).

Because of the intrinsic high computational cost to simulate fluid interfaces proper of fixed-grid Eulerian-based models, and the high distortions for violent fluid-structure interactions of the Lagrange method [46], both typical characteristics of UNDEX, the combination of the two into the Arbitrary Lagrangian Eulerian approach (ALE) is quite common in this field of study. It considers a moving grid to provide enhanced interfacial precision and to optimize the usage of computational resources for the calculations [47]. The high CPU cost to simulate fixed mesh problems has been addressed by several researchers. Wang et al. [48] were able to cut down this effect by defining optimal ratios between the radius of the charge and the side length of the mesh elements through a dimensionless variable, whose recommended value was found to range between 3 and 6 for most cases.

Even with its limitations for complex simulations, the description of compressible flow by Eulerian algorithms is still popular in literature. Liu et al. [49] developed a continuous UNDEX simulation model that embraces both shock wave generation and bubble motion stages consecutively to overcome the usual two-step routine. Through this method, the high-pressure bubbles become smaller and weaker than the real ones with errors up to 10%, most likely related to intense interaction of the reflection wave with the expanding bubble in the proximity of the wall. Hu et al. [50] managed to obtain satisfactory values for compressible multi-fluid flows prevenient from UNDEX (gas and liquid water) through a sharp-interface (i.e., considering discontinuous material properties across the interface). Accordingly, Ma et al. [51] successfully presented an extended version of a known set of one-dimensional equations for compressible multiphase fluids to two and three dimensions.

Purely Lagrangian techniques have also been effectively used to numerically reproduce UNDEX, such as in Ming et al. [18], where a faithful fracture pattern on metallic panels was achieved, and

Zhang et al. [52], where the penetration of metal jet on steel plates was accurate, thus, proving the interfacial efficiency of this method. Aiming to supply the faults and couple the benefits of both methods, ALE has also been applied by authors like Jafarian et al. [53] to allow the simulation of UNDEX compressible flow and cavitation (formation, development, and collapse of bubbles), using a single fluid by Petrov et al. [54] to take into account cavitation and rarefaction waves that propagate through the liquid, and by Wardlaw [55] to validate simulations at one, two, and three dimensions and with a variable number of moving boundaries.

## 3.2. Numerical Modelling

The Eulerian grid is applicable to solving methods like Finite Element Method (FEM) [2,28,30,31,36] and Finite Volume Method (FVM) [48,51,55,56], since both calculate the values of the fluid properties at discrete places on a meshed geometry. In FEM, these discretized spatial units are defined as elements and have constant properties, being mostly used to model solid ship structures and materials, although fluid domains can also be formulated [57]. Prusty et al. [31] and Gupta et al. [28] are good examples of the precision with which FEM models can reproduce the response of metallic stiffened panels to UNDEX. Also, FEM constitutes the best technique through which the hull girder strength can be assessed by a progressive collapse analysis and its consequent non-linearity [56], as Chung et al. [58] demonstrated by evaluating the structural response of a catamaran, even taking into account the strain rate effects of the material. Not confined to monolithic structures, Tillbrook et al. [37] demonstrated that beams made of sandwich material compositions can have their dynamic response to UNDEX adequately described by FEM, which is a recognized standard approach to model ship structures by naval classification agencies and the navy [59].

In turn, FVM supports unstructured meshes of discretized units named cells, through which matter is allowed to flow following conservative physical laws. Unlike in elements where boundary conditions may be applied in surfaces or nodes, in FVM these are directly applied in the control volume within the cell, thus not influencing neighbor matter units. Wang et al. [48] make use of FVM to define both water and explosive charge materials, allowing them to flow over the mesh cells, and Ma et al. [51] applied the FVM to model one, two, and three-dimensional problems with multiple fluids efficiently, proving the validity of this method. Although FVM is not a popular choice to model either structures or fluids, military-grade validations of its usage have been carried by studies such as Wardlaw's [55].

Other techniques with interesting resources, such as Boundary Element Method (BEM) [19,60–62] (also used for propeller hydrodynamic calculations [63]) and Smooth Particle Hydrodynamics (SPH) [18,52] have proven to be efficient, having their own intrinsic advantages.

BEM, also known as BIM (Boundary Integral Method), discretizes only the edges of the control volumes formed by a grid. Due to this grid simplicity, BEM allows the calculation of more complex integral equations with a reduced computational power, potentially resulting in highly accurate outputs. It is particularly popular to model problems involving bubble impact, as it provides efficient algorithms to represent its transient shape and pressure from the moment of the explosion until the collapse. Wang et al. [60] proposed numerical modifications to include not only the compressibility over the bubble to simulate its damped oscillation, but also the effect of oscillation cycles subsequent to the first one before it breaks down into smaller bubbles. Zhang et al. [61] introduced improvements to the regular 3D BEM model, enhancing the accuracy and stability at large deformations of the toroidal bubble phase. Li et al. [62] studied the pressure field caused by a collapsing bubble by the auxiliary function method, finding out that its dynamic total pressure can be decomposed and quantified in two parts: one regarding the pressure gradient between the gas within the bubble and surrounding liquid, and another related to its motion. Coupled approaches have also been explored for hydrodynamic problems, such as RANS-BEM (Reynolds-averaged Navier-Stokes) [20] and BEM-FEM. For example, the latter, as studied by Gong et al. [19], allowed the simultaneous simulation of the physics of bubble growth, contraction, and collapse, and the consequent behavior of the hull. The shape transformation of the bubble from its generation until impact, as generically represented in these studies, is displayed in Figure 6.

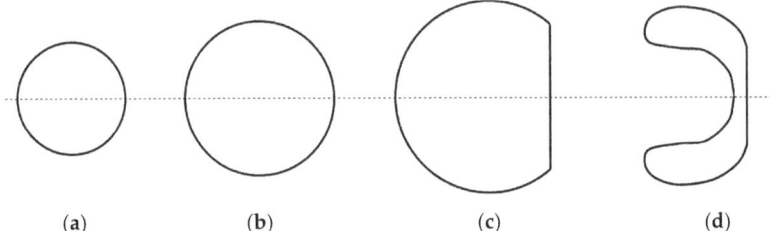

**Figure 6.** Representation of steps of bubble mid-sectional shape change starting from its conception (**a**), expansion due to increase in inner gas pressure (**b**), flattening on rigid surface and attainment of maximum volume (**c**), to the collapsing toroidal shape (**d**).

However, the discretization of a domain in elements might lead to difficulties to deal with free-surfaces, deformable boundaries, moving interfaces, and large fluid deformations, and assuring a satisfactory mesh quality may often be a tricky and time-consuming process. Within this context, Lagrangian based methods, such as SPH, present an interesting alternative to overcome the barriers imposed by a grid system [64]. SPH is a robust method for intricate hydrodynamic problems [46] that represents a set of particles with material properties that interact with each other in the flow. This method is attributed to numerical simulations, in which the analysis deals with large deformations, given that it has a meshless nature. Zhang et al. [52] applied SPH to simulate a charge detonation, the metal-jet formation, and the penetration on a steel plate, agreeing with experimental results. Although impacts on the air-water interface constitute a limitation of the SPH-FEM coupling because of its instabilities in modelling particular water physics [65], it is a valid and precise approach for UNDEX. Ming et al. [18] demonstrated through the comparison of experiments using SPH and SPH-FEM methods that both are accurate to reproduce UNDEX. De Vuyst et al. [66] have actually proved through coupled SPH-FEM simulations the interesting concept that considering the case of having either one or multiple underwater explosions for a same charge quantity, the latter is less harmful to the steel than the first, as it generally results in smaller final deformation and might avoid cracks that could be formed by a single blast.

Given the aforementioned considerable amount of input on numerical modelling techniques, a similar approach to Hirdaris et al. [67] is reflected in Table 1 to summarize the key features of each method addressed in this subsection, aiming to provide a clearer visualization of their particularities. The taxonomy for classification adopted is based on the level of mesh dependency they present to yield accurate outputs, as some methods always include discretization, others partially, and others none.

It is vital to state that a taxonomic descriptive table was preferred to point out the peculiarities of each method, because it is not possible to indicate in a simple manner which method is more appropriate in each occasion; after all, the method choice is subjective for the user, and takes into account their affinity with specific software, the CPU power available, the level of precision requested, and the boundary conditions of the problem to be analyzed. Hence, the same method can present diverse levels of complexity depending on the situation. Besides, through the scientific review hereby carried, it was possible to notice that all methods are capable of providing valid results. Despite this, the survey showed a literature inclination to FEM to reproduce vessel structures, and to BEM and SPH to model the fluids, be they gas (prevenient from bubbles), liquid, or both.

Overcoming the academic sphere, naval classification agencies and official military reports also support these methods, confirming their legitimacy to simulate UNDEX. Holtmann from Det Norske Veritas [59] advises FEM to evaluate the shock resistance, accelerations, strains, and stresses of hull girders for providing a fast and efficient approach. Wang et al. from American Bureau of Shipping make use of nonlinear FEM to design ship structures for ice loads [68]. Warlaw [55] from the U.S. Navy considers the division of the UNDEX simulation domain into FVM cells applied to problems with different dimensions and border conditions. Det Norske Veritas [69] sees BEM as the most common method to solve potential flow problems, whereas O'Daniel et al. [16] from the U.S. Army faces this method as the most suitable to simulate the behavior of bubbles. Also, Jones et al. [70] from the Department of Defense of Australia recognizes SPH as appropriate for modeling several fluid problems of particular military interest.

**Table 1.** Taxonomy of hydrodynamic modelling methods as a function of their discretization dependency.

| Level | Method | Key Features | References |
|---|---|---|---|
| 1 | FEM | <ul><li>Mesh-dependent</li><li>Conventionally used to simulate the behavior of materials and ship structures</li><li>Can be used to model nonlinear impact</li><li>Impractical for violent flows</li><li>If large deformations are considered, the mesh must be too fine and computations become slow</li></ul> | Latourte et al. [2]<br>Gupta et al. [28]<br>Hung et al. [30]<br>Prusty et al. [31]<br>Tillbrook et al. [37]<br>Holtmann et al. [59]<br>Wang et al. [68] |
| 2 | FVM | <ul><li>Mesh dependent, but allows unstructured meshes</li><li>Cells obey conservative laws</li><li>Boundary conditions are applied noninvasively within the volume cell</li></ul> | Wang et al. [48]<br>Ma et al. [51]<br>Wardlaw [55]<br>Rigo et al. [56] |
| 3 | BEM | <ul><li>It is highly accurate due to the simplicity of the grid that allows the adoption of complex integral equations</li><li>Velocity potential in fluid domain represented by a distribution of sources over the mean wet body surface</li><li>Suitable for bubble formation, growth and collapse</li><li>Precise for problems involving stress concentrations</li><li>May give unreliable values for added mass and damping at irregular shock frequencies</li></ul> | O'Daniel et al. [16]<br>Gong et al. [19]<br>Wang [60]<br>Zhang et al. [61]<br>Li et al. [62]<br>DNV [69] |
| 4 | SPH | <ul><li>Meshless</li><li>The Lagrangian nature tracks the mass of material particles and difficulties boundary conditions setup</li><li>Advection can be calculated</li><li>Adequate for free-surface, interfacial and violent flow</li><li>Exact and simultaneous conservation of mass, momentum, angular momentum, energy and entropy</li><li>Computationally expensive due to higher number of neighbor elements</li><li>Easy to parallelize. So, even if it is costlier than other methods, it has the potential to be faster</li></ul> | Ming et al. [18]<br>Zhang et al. [46]<br>Zhang et al. [52]<br>Liu et al. [64]<br>Hughes et al. [65]<br>De Vuyst et al. [66]<br>Jones et al. [70] |

*3.3. Constitutive Models*

The techniques used to model impact solicitations differ depending on the type of material considered [71–73], and particular sets of equations must be adopted to form a software-embedded constitutive model, which generally requires calibration of variables. Often, given the wide application of metals in shipbuilding, the models used for UNDEX make use of relations developed for metallic materials, which are generally highly dependent on strain rate effects [74].

Accordingly, ship design relies on such resource to properly characterize the vessel itself and all its structural components to efforts intrinsic of navigation, as well as to external input forces such as UNDEX. A variety of studies deal with different strategies to model the aforementioned problem, exhibiting the versatility with which it can be done by using diverse approaches and modeling techniques.

Characterizing a common ground among most authors, the application of the Johnson-Cook constitutive model [75] for the ship hull or single panels is widely utilized, coupled with equations of state in numerical codes, allowing its usage for UNDEX loading. Furthermore, it accounts for equivalent plastic strain, strain rate, and the influence of temperature, while requiring only five parameters that can be found by means of experimental tests, as described in Equation (23):

$$\sigma = \left(A + B\varepsilon_{eff}^n\right)\left(1 + Cln\varepsilon_{ref}\right)\left(1 - T_h^m\right) \qquad (23)$$

where $\varepsilon_{eff}$ is the effective plastic strain, $\varepsilon_{ref}$ is the same strain but at a reference rate $\varepsilon_0 = 1s^{-1}$ and $T_h$ is the homologous temperature calculated by the quotient of the difference between the material temperature and the room temperature by the difference between the melting temperature and the room temperature. As for the five parameters, namely $A$, $B$, $C$, $m$, and $n$, the first two and the latter are obtained from tension tests, while $C$ and $m$ are defined through split Hopkinson pressure bar tests. This model was used by Kong et al. [76], which numerically reproduced a blast load detonated inside a multi-layer protective and stiffened naval structure, demonstrating in this case that stiffened plates are severely damaged by fragments which penetrate, causing diverse spots of crack initiation culminating into crack propagation. On relatable UNDEX studies [18,28], stiffened plates were also modeled, making use of the Johnson-Cook criterion.

Another relevant constitutive model used in this field, although secondary compared to Johnson-Cook's for being less accurate to represent UNDEX, is Gurson's [77]. It can be calibrated by the realization of mechanical tests [2] and describes stress flow and rupture of materials depending on void growth. A description of the damage growth rate ($\dot{f}$) as a function of the void volume fraction ($f$) and the plastic strain rate ($D^P$) in a pure shear stress case (i.e., in the $kk$ direction) follows (Equation (24)):

$$\dot{f} = (1-f)D_{kk}^P \qquad (24)$$

In this case, the limitation assumed by this model is that no damage is predicted when strain is under zero mean stress, depending exclusively on the formation of voids. Aiming to address this issue and include the effect of shear-induced fracture also in the absence of voids, a modification of this model was introduced by Nahshon et al. [78], which adds the numerical constant $k_\omega$ to define the rate of damage evolution in a shear-predominant stress state. The evolution of the model is thus presented by the incorporation of a second term on Equation (24) resulting in Equation (25):

$$\dot{f} = (1-f)D_{kk}^P + k_\omega f \omega(\sigma)\frac{s_{ij}D_{ij}^P}{\sigma_e} \qquad (25)$$

where $\omega(\sigma)$ describes the relation between a third invariant of stress with effective stress [78], $s_{ij}$ is the stress deviator $s_{ij} = \sigma_{ij} - 1/3\sigma_{kk}\delta_{ij}$, and $\sigma_e$ stands for the effective stress. This enhanced approach was validated by Xue et al. [79], where the calibration of the damage parameters inherent of this constitutive model were able to reproduce tension and shear induced failures on steel coupons. Latourte et al. [2] used it to numerically describe the behavior of high performance steel alloys to UNDEX.

As for sandwich panels, studies such as the ones from Tilbrook et al. [37,80], considering anisotropic foam cores and elastically ideal skin sheets, provide particular constitutive models allying the mechanics of core compression with the bending response of sandwich beams. These two impulsive load responses can be described in four ways, depending on whether they are analyzed separately (if the accelerations of the sheets differ) or together, and if the core densification is total or partial.

## 4. Discussion

Addressing the mechanics embraced by underwater explosions, the present survey underlines essential mathematical relations from the hydrodynamics of subsea explosions, to the propagation of high pressure waves and bubbles, to the behavior shown by materials and structures commonly used in shipbuilding. The validity of these methods has been checked through the high correlation of purely analytical formulations with several experimental works, although some discrepancies among authors have also been identified, such as the constants used for determining peak pressure as a function of the charge load.

Likewise, the variety of computational numerical resources applied in this area have permitted the constitution of advanced research studies that are able to mirror UNDEX, conditions considering complex wave force transmission mechanisms and progressively accurate material responses in parallel, thus ratifying the forefront role played by virtual simulations in related engineering problems.

Moreover, the confrontation of the late efforts in improving and perfecting the precise reproducibility of constitutive models to real situations, with the reported advanced material characterization and hydrodynamic studies, show that this traditional but ongoing research field has kept its evolving pace and still presents a remarkable development potential.

Among the possible future directions for this area of study, it is hereby encouraged that a comparative study considering different numerical simulation techniques (such as FVM, BEM, and SPH) is carried, adopting the same boundary conditions, aiming to narrow the current subjectivity in selecting the modelling method to a more grounded approach, crossing desired output accuracy with computational power available. Regarding the analytical side, it would be considerably relevant if new experimental studies could confront the response of materials for hull panels to UNDEX to those in the sometimes-conflicting studies available in the literature. In this case, special attention should be given to fiber reinforced composite materials for their growing importance in the nautical industry over the past years, considering their behavior to underwater explosions has not yet been properly explored.

**Funding:** This research received no external funding.

**Conflicts of Interest:** The author declares no conflict of interest.

## References

1. Pedersen, P.T.; Zhang, S. On impact mechanics in ship collisions. *Mar. Struct.* **1998**, *11*, 429–449. [CrossRef]
2. Latourte, F.; Wei, X.; Feinberg, Z.D.; De Vaucorbeil, A.; Tran, P.; Olson, G.B.; Espinosa, H.D. Design and identification of high performance steel alloys for structures subjected to underwater impulsive loading. *Int. J. Solids Struct.* **2012**, *49*, 1573–1587. [CrossRef]
3. Fragassa, C.; Minak, G. Measuring Deformations in a Rigid-Hulled Inflatable Boat. *Key Eng. Mater.* **2017**, *754*, 295–298. [CrossRef]
4. Liang, Y.; Spuskanyuk, A.V.; Flores, S.E.; Hayhurst, D.R.; Hutchinson, J.W.; McMeeking, R.M.; Evans, A.G. The response of metallic sandwich panels to water blast. *J. Appl. Mech.* **2007**, *74*, 81–99. [CrossRef]
5. Fragassa, C. From Design to Production: An integrated advanced methodology to speed up the industrialization of wooden boats. *J. Ship Prod. Des.* **2017**, *33*, 237–246. [CrossRef]
6. Fragassa, C. Marine Applications of Natural Fibre-Reinforced Composites: A Manufacturing Case Study. In *Advances in Application of Industrial Biomaterials*; Pellicer, E., Nikolic, D., Sort, J., Baró, M., Zivic, F., Grujovic, N., Grujic, R., Pelemis, S., Eds.; Springer International Publishing: Cham, Switzerland, 2017; pp. 21–47. ISBN 978-3-319-62766-3.
7. Lorenzini, G.; Helbig, D.; Real, M.V.; Dos Santos, E.D.; Isoldi, L.A.; Rocha, L.A.O. Computational modeling and constructal design method applied to the mechanical behavior improvement of thin perforated steel plates subject to buckling. *J. Eng. Thermophys.* **2016**, *25*, 197–215. [CrossRef]
8. Lorenzini, G.; Helbig, D.; Da Silva, C.C.C.; Real, M.V.; Dos Santos, E.D.; Isoldi, L.A.; Rocha, L.A.O. Numerical evaluation of the effect of type and shape of perforations on the buckling of thin steel plates by means of the constructal design method. *Heat Tech.* **2016**, *34*, S9–S20. [CrossRef]
9. Helbig, D.; Da Silva, C.C.C.; Real, M.V.; Dos Santos, E.D.; Isoldi, L.A.; Rocha, L.A.O. Study About Buckling Phenomenon in Perforated Thin Steel Plates Employing Computational Modeling and Constructal Design Method. *Latin Am. J. Solids Struct.* **2016**, *13*, 1912–1936. [CrossRef]
10. Keil, A.H.; UERD, Norfolk Naval Ship Yard, Portsmouth, VA, USA. Introduction to underwater explosion research, 1956.
11. Johnson, W.; Poyton, A.; Singh, H.; Travis, F.W. Experiments in the underwater explosion stretch forming of clamped circular blanks. *Int. J. Mech. Sci.* **1966**, *8*, 237–270. [CrossRef]

12. Strahle, W.C.; Georgia Institute of Technology, Atlanta, GA, USA. Investigation of research needs for underwater explosions, 1990.
13. Barash, R.M.; United States Naval Ordnance Laboratory, White Oak, MD, USA. Underwater explosions beneath ice, 1962.
14. Bryant, E.F.; Malaker Laboratories Inc., High Brigde, NJ, USA. Debris distribution in underwater explosions, 1964.
15. Snay, H.G.; United States Naval Ordnance Laboratory, White Oak, MD, USA. Hydrodynamic concepts selected topics for underwater nuclear explosions, 1966.
16. O'Daniel, J.L.; Harris, G.; Ilamni, R.; Chahine, G.; Fortune, J. *Underwater Explosion Bubble Jetting Effects on Infrastructure*; US Army Corps of Engineers—ERDC Vicksburg: Vicksburg, MS, USA, 2011.
17. Miller, W.E. Simulation of the Underwater Nuclear Explosion and Its Effects. Ph.D. Thesis, Naval Postgraduate School, Monterey, CA, USA, June 1992.
18. Ming, F.R.; Zhang, A.M.; Xue, Y.Z.; Wang, S.P. Damage characteristics of ship structures subjected to shockwaves of underwater contact explosions. *Ocean Eng.* **2016**, *117*, 359–382. [CrossRef]
19. Gong, S.W.; Khoo, B.C. Transient response of stiffened composite submersible hull to underwater explosion bubble. *Compos. Struct.* **2015**, *122*, 229–238. [CrossRef]
20. Regener, P.B.; Mirsadraee, Y.; Andersen, P. Nominal vs. Effective Wake Fields and their Influence on Propeller Cavitation Performance. *J. Mar. Sci. Eng.* **2018**, *6*, 34. [CrossRef]
21. Brenner, M. *Navy Ship Underwater Shock Prediction and Testing Capability Study*; Report-No. JSR, 07-200; MITRE Corporation: McLean, VA, USA, 2007.
22. De Vuyst, T.; Vignjevic, R.; Albero, A.A.; Hughes, K.; Campbell, J.C.; Djordjevic, N. The effect of the orientation of cubical projectiles on the ballistic limit and failure mode of AA2024-T351 sheets. *Int. J. Impact Eng.* **2017**, *104*, 21–37. [CrossRef]
23. Cole, R.H.; Weller, R. Underwater explosions. *Phys. Today* **1948**, *1*, 35. [CrossRef]
24. Cui, P.; Zhang, A.M.; Wang, S.P. Small-charge underwater explosion bubble experiments under various boundary conditions. *Phys. Fluids* **2016**, *28*, 117103. [CrossRef]
25. Zamyshlyayev, B.V.; Yakovlev, Y.S. *Dynamic Loads in Underwater Explosion*; Naval Intelligence Support Center: Washington, DC, USA, 1973.
26. Keil, A.H. *The Response of Ships to Underwater Explosions*; Report No. DTMB-1576; David Taylor Model Basin: Washington, DC, USA, 1961.
27. Rajendran, R.; Narasimhan, K. Deformation and fracture behaviour of plate specimens subjected to underwater explosion—A review. *Int. J. Impact Eng.* **2006**, *32*, 1945–1963. [CrossRef]
28. Gupta, N.K.; Kumar, P.; Hegde, S. On deformation and tearing of stiffened and un-stiffened square plates subjected to underwater explosion—A numerical study. *Int. J. Mech. Sci.* **2010**, *52*, 733–744. [CrossRef]
29. Rajendran, R.; Paik, J.K.; Kim, B.J. Design of warship plates against underwater explosions. *Ships Offshore Struct.* **2006**, *1*, 347–356. [CrossRef]
30. Hung, C.F.; Lin, B.J.; Hwang-Fuu, J.J.; Hsu, P.Y. Dynamic response of cylindrical shell structures subjected to underwater explosion. *Ocean Eng.* **2009**, *36*, 564–577. [CrossRef]
31. Prusty, B.G.; Satsangi, S.K. Analysis of stiffened shell for ships and ocean structures by finite element method. *Ocean Eng.* **2001**, *28*, 621–638. [CrossRef]
32. Gordo, J.M.; Soares, C.G.; Faulkner, D. Approximate assessment of the ultimate longitudinal strength of the hull girder. *J. Ship Res.* **1996**, *40*, 60–69.
33. Gordo, J.M.; Soares, C.G. Approximate load shortening curves for stiffened plates under uniaxial compression. *Integr. Offshore Struct.* **1993**, *5*, 189–211.
34. Chen, Y.; Tong, Z.P.; Hua, H.X.; Wang, Y.; Gou, H.Y. Experimental investigation on the dynamic response of scaled ship model with rubber sandwich coatings subjected to underwater explosion. *Int. J. Impact Eng.* **2009**, *36*, 318–328. [CrossRef]
35. Watson, D.G.M. *Practical Ship Design*; Elsevier: Amsterdam, The Netherlands, 1998; Volume 1, ISBN 978-0-0804-2999-1.
36. Tupper, E.C.; Rawson, K.J. *Basic Ship Theory*; Butterworth-Heinemann: Oxford, UK, 2001; Volume 2, ISBN 978-0-7506-5398-5.
37. Tilbrook, M.T.; Deshpande, V.S.; Fleck, N.A. Underwater blast loading of sandwich beams: Regimes of behaviour. *Int. J. Solids Struct.* **2009**, *46*, 3209–3221. [CrossRef]

38. Fleck, N.A.; Deshpande, V.S. The resistance of clamped sandwich beams to shock loading. *J. Appl. Mech.* **2004**, *71*, 386–401. [CrossRef]
39. Hutchinson, J.W.; Xue, Z. Metal sandwich plates optimized for pressure impulses. *Int. J. Mech. Sci.* **2005**, *47*, 545–569. [CrossRef]
40. Fan, Z.; Liu, Y.; Xu, P. Blast resistance of metallic sandwich panels subjected to proximity underwater explosion. *Int. J. Impact Eng.* **2016**, *93*, 128–135. [CrossRef]
41. Rajendran, R.; Narasimhan, K. Linear elastic shock response of plane plates subjected to underwater explosion. *Int. J. Impact Eng.* **2001**, *25*, 493–506. [CrossRef]
42. Rajendran, R.; Narasimhan, K. Underwater shock response of circular HSLA steel plates. *Shock Vibr.* **2000**, *7*, 251–262. [CrossRef]
43. Rajendran, R.; Narasimhan, K. Damage prediction of clamped circular plates subjected to contact underwater explosion. *Int. J. Impact Eng.* **2001**, *25*, 373–386. [CrossRef]
44. Jones, N. *Structural Impact*, 2nd ed.; Cambridge University Press: Cambridge, UK, 2012; ISBN 978-1-1070-1096-3.
45. Jacob, N.; Yuen, S.C.K.; Nurick, G.N.; Bonorchis, D.; Desai, S.A.; Tait, D. Scaling aspects of quadrangular plates subjected to localised blast loads—Experiments and predictions. *Int. J. Imp. Eng.* **2004**, *30*, 1179–1208. [CrossRef]
46. Zhang, A.M.; Sun, P.N.; Ming, F.R.; Colagrossi, A. Smoothed particle hydrodynamics and its applications in fluid-structure interactions. *J. Hydrodyn.* **2017**, *29*, 187–216. [CrossRef]
47. Helenbrook, B.T.; Hrdina, J. High-order adaptive arbitrary-Lagrangian–Eulerian (ALE) simulations of solidification. *Comput. Fluids* **2018**, *167*, 40–50. [CrossRef]
48. Wang, G.; Wang, Y.; Lu, W.; Zhou, W.; Chen, M.; Yan, P. On the determination of the mesh size for numerical simulations of shock wave propagation in near field underwater explosion. *Appl. Ocean Res.* **2016**, *59*, 1–9. [CrossRef]
49. Liu, W.T.; Ming, F.R.; Zhang, A.M.; Miao, X.H.; Liu, Y.L. Continuous simulation of the whole process of underwater explosion based on Eulerian finite element approach. *Appl. Ocean Res.* **2018**, *80*, 125–135. [CrossRef]
50. Hu, X.Y.; Adams, N.A.; Iaccarino, G. On the HLLC Riemann solver for interface interaction in compressible multi-fluid flow. *J. Comput. Phys.* **2009**, *228*, 6572–6589. [CrossRef]
51. Ma, Z.H.; Causon, D.M.; Qian, L.; Gu, H.B.; Mingham, C.G.; Ferrer, P.M. A GPU based compressible multiphase hydrocode for modelling violent hydrodynamic impact problems. *Comput. Fluids* **2015**, *120*, 1–23. [CrossRef]
52. Zhang, Z.; Wang, L.; Silberschmidt, V.V. Damage response of steel plate to underwater explosion: Effect of shaped charge liner. *Int. J. Impact Eng.* **2017**, *103*, 38–49. [CrossRef]
53. Jafarian, A.; Pishevar, A. An exact multiphase Riemann solver for compressible cavitating flows. *Int. J. Multiphase Flow* **2017**, *88*, 152–166. [CrossRef]
54. Petrov, N.V.; Schmidt, A.A. Multiphase phenomena in underwater explosion. *Exp. Therm. Fluid Sci.* **2015**, *60*, 367–373. [CrossRef]
55. Wardlaw, A.B., Jr. *Underwater Explosion Test Cases*; No. NSWC-IHTR-2069; Naval Surface Warfare Center: Indian Head, MD, USA, 1998.
56. Rigo, P.; Rizzuto, E. Analysis and Design of Ship Structure. *Ship Des. Constr.* **2003**, *1*, 18-1.
57. Zhang, W.; Yao, X.; Liu, L.; Wang, Z. Semi-analytical and experimental investigation of the whipping response of a cylinder subjected to underwater explosion load. *Ships Offshore Struct.* **2018**, 1–9. [CrossRef]
58. Chung, J.; Shin, Y.S. Simulation of dynamic behaviour of high-speed catamaran craft subjected to underwater explosion. *Ships Offshore Struct.* **2013**, *9*, 387–403. [CrossRef]
59. Det Norske Veritas. Available online: https://www.dnvgl.com/services/shock-analysis-4716 (accessed on 30 November 2018).
60. Wang, Q. Multi-oscillations of a bubble in a compressible liquid near a rigid boundary. *J. Fluid Mech.* **2014**, *745*, 509–536. [CrossRef]
61. Zhang, A.M.; Liu, Y.L. Improved three-dimensional bubble dynamics model based on boundary element method. *J. Comput. Phys.* **2015**, *294*, 208–223. [CrossRef]
62. Li, S.; Han, R.; Zhang, A.M.; Wang, Q.X. Analysis of pressure field generated by a collapsing bubble. *Ocean Eng.* **2016**, *117*, 22–38. [CrossRef]

63. Maljaars, P.; Kaminski, M.; den Besten, H. Boundary element modelling aspects for the hydro-elastic analysis of flexible marine propellers. *J. Mar. Sci. Eng.* **2018**, *6*, 67. [CrossRef]
64. Liu, M.B.; Liu, G.R. Smoothed particle hydrodynamics (SPH): An overview and recent developments. *Arch. Comput. Methods Eng.* **2010**, *17*, 25–76. [CrossRef]
65. Hughes, K.; Vignjevic, R.; Campbell, J.; De Vuyst, T.; Djordjevic, N.; Papagiannis, L. From aerospace to offshore: Bridging the numerical simulation gaps–Simulation advancements for fluid structure interaction problems. *Int. J. Impact Eng.* **2013**, *61*, 48–63. [CrossRef]
66. De Vuyst, T.; Kong, K.; Djordjevic, N.; Vignjevic, R.; Campbell, J.C.; Hughes, K. Numerical modelling of the effect of using multi-explosives on the explosive forming of steel cones. *J. Phys. Conf. Ser.* **2016**, *734*, 032074. [CrossRef]
67. Hirdaris, S.E.; Lee, Y.; Mortola, G.; Incecik, A.; Turan, O.; Hong, S.Y.; Kim, B.W.; Kim, K.H.; Bennett, S.; Miao, S.H.; et al. The influence of nonlinearities on the symmetric hydrodynamic response of a 10,000 TEU Container ship. *Ocean Eng.* **2016**, *111*, 166–178. [CrossRef]
68. Wang, G.; Wiernicki, C.J. Using nonlinear finite element method to design ship structures for ice loads. *Mar. Tech.* **2006**, *43*, 1–15.
69. Det Norske Veritas. *Recommended Practice DNV-RP-C205: Environmental Conditions and Environmental Loads*; Det Norske Veritas: Hovik, Norway, 2014.
70. Jones, D.A.; Belton, D. *Smoothed Particle Hydrodynamics: Applications within DSTO (No. DSTO-TR-1922)*; Defence Sci. Tech. Org.; Platform Sciences Lab: Fishermans Bend, Victoria, Australia, 2006.
71. Fragassa, C.; Camargo, F.V.; Pavlovic, A.; Minak, G. Explicit numerical modeling assessment of basalt reinforced composites for low-velocity impact. *Comp. Part B: Eng.* **2019**, in press. [CrossRef]
72. Pavlovic, A.; Camargo, F.V.; Fragassa, C. Crash safety design: Basic principles of impact numerical simulations for composite materials. In Proceedings of the 9th International Conference on Times of Polymers and Composites (AIP Conference Proceedings), Ischia, Italy, 17–21 June 2018; D'Amore, A., Grassia, L., Acierno, D., Eds.; AIP Publishing: Melville, NY, USA, 2018; Volume 1981, p. 020032. [CrossRef]
73. Fragassa, C.; Camargo, F.V.; Pavlovic, A.; Silveira, A.C.F.; Minak, G.; Bergmann, C.P. Mechanical Characterization of Gres Porcelain and Low-Velocity Impact Numerical Modelling. *Materials* **2018**, *11*, 1082. [CrossRef] [PubMed]
74. Djordjevic, N.; Vignjevic, R.; Kiely, L.; Case, S.; De Vuyst, T.; Campbell, J.; Hughes, K. Modelling of shock waves in fcc and bcc metals using a combined continuum and dislocation kinetic approach. *Int. J. Plast.* **2018**, *105*, 211–224. [CrossRef]
75. Johnson, G.R.; Cook, W.H. A constitutive model and data for metals subjected to large strain, high strain rates and high temperatures. In Proceedings of the 7th International Symposium on Ballistics, The Hague, The Netherlands, 19–21 April 1983; pp. 541–547.
76. Kong, X.S.; Wu, W.G.; Li, J.; Chen, P.; Liu, F. Experimental and numerical investigation on a multi-layer protective structure under the synergistic effect of blast and fragment loadings. *Int. J. Impact Eng.* **2014**, *65*, 146–162. [CrossRef]
77. Gurson, A.L. Continuum theory of ductile rupture by void nucleation and growth: Part I—Yield criteria and flow rules for porous ductile media. *J. Eng. Mat. Tech.* **1977**, *99*, 2–15. [CrossRef]
78. Nahshon, K.; Hutchinson, J.W. Modification of the Gurson model for shear failure. *Eur. J. Mech. A/Solids* **2008**, *27*, 1–17. [CrossRef]
79. Xue, Z.; Pontin, M.G.; Zok, F.W.; Hutchinson, J.W. Calibration procedures for a computational model of ductile fracture. *Eng. Fract. Mech.* **2010**, *77*, 492–509. [CrossRef]
80. Tilbrook, M.T.; Deshpande, V.S.; Fleck, N.A. Regimes of response for impulse loaded sandwich panels. *J. Mech. Phys. Solids* **2006**, *54*, 2242–2280. [CrossRef]

© 2019 by the author. Licensee MDPI, Basel, Switzerland. This article is an open access article distributed under the terms and conditions of the Creative Commons Attribution (CC BY) license (http://creativecommons.org/licenses/by/4.0/).

Article

# An Approach for Predicting the Specific Fuel Consumption of Dual-Fuel Two-Stroke Marine Engines

Crístofer H. Marques [1,*], Jean-D. Caprace [2], Carlos R. P. Belchior [2] and Alberto Martini [3]

1. School of Engineering, Federal University of Rio Grande, Rio Grande RS 96203-900, Brazil
2. Ocean and Naval Engineering Department, Federal University of Rio de Janeiro, Rio de Janeiro RJ 21941-901, Brazil; jdcaprace@oceanica.ufrj.br (J.-D.C.); belchior@oceanica.ufrj.br (C.R.P.B.)
3. Department of Industrial Engineering, University of Bologna, Viale del Risorgimento 2, 40126 Bologna, Italy; alberto.martini6@unibo.it
* Correspondence: cristoferhood@furg.br

Received: 4 October 2018; Accepted: 9 January 2019; Published: 22 January 2019

**Abstract:** Increasing environmental demands, alongside the planned penetration of natural gas as marine fuel, have rendered dual-fuel engines as an attractive prime mover alternative. In this context, knowing the specific fuel consumption is essential to selecting the most efficient engine. The specific fuel consumption can be approached by simulation models with varying levels of complexity that are either implemented by basic programming languages or simulated by dedicated packages. This study aims to develop a simplified model to predict the specific fuel consumption of dual-fuel two-stroke marine engines driving fixed or controllable pitch propellers. The model relies on clear trends approachable by polynomials that were revealed by normalizing specific fuel consumption. This model requires only the value of specific fuel consumption at a nominal maximum continuous rating to predict the engine consumption at any specified rating, including at partial engine load. The outcome of the study shows that the maximum deviations regarding the two simulated engines did not exceed −3.6%. In summary, the proposed model is a fast and effective tool for optimizing the selection of dual-fuel, two-stroke Diesel engines regarding fuel consumption.

**Keywords:** fixed pitch propeller; controllable pitch propeller; low-speed Diesel engine; selection; optimisation; modelling

---

## 1. Introduction

The maritime industry has faced new realities that have been changing marine fuel investment choices over the last few decades. Although vessels have become cleaner, regulators, environmentalists, and health officials have still been concerned about pollutants near major coastal population centres [1]. Furthermore, the decision to implement a global sulphur cap of 0.5% in 2020, revising the current 3.5% cap (outside sulphur emission control areas), was presented in the Resolution MEPC.176(58) [2] and confirmed by the International Maritime Organisation (IMO) on October 2016 [3]. This change applies globally and will affect as many as 70,000 ships, which is a reason why experts do not agree completely with the IMO's study that indicates the refineries will be capable of providing the required amount of low-sulphur marine fuel by 2020 [4].

Natural gas offers lower local pollution emissions compared to distillate fuels, and can significantly reduce local pollutants from vessel operations. Price differences between natural gas and low-sulphur fuel oil suggest that an economic advantage may favor the use of natural gas. In addition, natural gas infrastructure has been growing, rendering ships fed by natural gas more plausible [1]. These have been some of the reasons why dual-fuel Diesel engines have become an attractive prime mover alternative.

The term "dual-fuel" describes compression ignition engines burning two different fuels simultaneously in varying proportions. In gas mode, gaseous fuel supplies most of the energy released through combustion, whereas liquid fuel is employed to provide the energy needed for ignition [5]. Hence, in this operation mode, there are two specific fuel consumptions: specific gas consumption (SGC), and specific pilot oil consumption (SPOC). In Diesel mode, these engines work as a conventional Diesel engine, such that there is only specific fuel oil consumption (SFOC).

Knowing these parameters is essential for selecting the most efficient engine and estimating its gaseous emission. The fuel consumption is the primary driver for operational expenditures, and it is directly linked to carbon dioxide emission, which is one of the greenhouse gases. Thus, in an optimization study, it is sought to select the engine of the least specific fuel consumption. Moreover, predicting the amount of fuel to be consumed in a journey is essential to design the fuel supply system. These are some of the reasons why the prediction of specific fuel consumption is relevant and addressed herein.

Internal combustion engine simulation consists in reproducing mathematically the significant processes, and predicting performance and operating details. The mathematical formulation for this purpose may be implemented through many scientific languages, such as Fortran, MatLab, GNU Octave, Scilab, C#, and C++. Some simulation-dedicated packages may also be applied, such as CORAL, CSMP, ACSL, Xcos, and SIMULINK. Furthermore, dedicated software, such as AVL BOOST, GT-POWER, and VIRTUAL ENGINE, may be applied for engine 0-D simulations, whilst multi-dimensional simulations may be performed through CONVERGE, KIVA, FLUENT, CFX, OpenFOAM, ANSYS Forte, and others [6].

The five main sorts of engine models in descending order of complexity are: computational fluid dynamics (CFD) models, phenomenological multi-dimensional models, crank angle models, mean value models, and transfer function models [7]. In CFD modelling, the volume studied is divided into thousands of parts, and the basic conservation equations are solved for every single part. This provides detailed information and requires powerful computers and high computational time. On the other hand, if the cylinder is divided into tens of volumes and phenomenological equations are included, a phenomenological multi-dimensional model is obtained. Crank angle models are also called zero-dimensional (0-D) because they do not have a strict mathematical dependence on any of the dimensions. It treats each of the various engine elements as a control volume and solves the differential equations in a time-step equivalent to one degree of the crankshaft rotation. Nevertheless, once the engine model is inserted into a larger system, such as a propulsion system, the variations occurring at each crankshaft angle are generally not important. In such cases, overall engine operating parameters are the focus, and they can be obtained by using a mean-value engine model (MVEM). This model has basically the same origin as the 0-D, but as its time-step is in the order of one crankshaft rotation, the variation of each parameter in the cylinder is replaced by a mean value. Finally, once there is no interest at all in the internal processes, the engine can be merely represented by functions. This is the so-called transfer function engine model (TFEM), which is the fastest method.

As the development of marine Diesel engines is a time-consuming and costly procedure, detailed engine-modelling techniques have been used for investigating the engine's steady-state performance and transient response, as well as for testing the alternative designs of the engine systems. Performance under fault conditions [8], the formation of noxious emissions [9–11], the employment of exhaust gas recirculation to decrease it [12], and instantaneous change in the properties of exhaust gas [13] besides the condensation of combustion products [14] may be assessed by 0-D modelling. On the other hand, the simulation of in-cylinder flow on different piston-bowl geometries [15], the investigation of soot formation and oxidation processes [16], as well as the investigation of the effects of inlet pressure, exhaust-gas recirculation, and the start of injection time on gaseous emissions [17], may be performed by CFD modelling.

In certain types of systems, a large number of components and feasible alternatives regarding design specifications and operating conditions makes the use of simulation and optimization techniques rather imperative to find a technical and economical attractive solution. In this context, dynamic simulations of the propulsion plant were explored in [18–21] whilst the synthesis, design, and/or operation optimization was addressed in [22–26]. In the majority of the works, the engine was modelled by means of the TFEM [18,21,23–27] and the MVEM was employed in a few of them [19,20]. This is due to the ease and speed of those models, as well as to the usual disinterest in in-cylinder parameters for engines inserted in larger systems. The simplest approach considered the prime mover as a constant value of specific fuel consumption [22]. Two-stroke Diesel engines were employed in [19,20,23] whilst a dual-fuel two-stroke Diesel engine was used in only one study [26].

In reference [26], specific fuel consumptions (SFOC, SPOC, and SGC) were modelled as functions of the specified maximum continuous rating (SMCR) and engine load, which is a fraction of the SMCR. The analysis was conducted by applying polynomial regression and the least square fitting method on data identified in the Project Guide of the engine's manufacturer [28]. However, specific fuel consumptions for the same SMCR are different to different engines and vary with the engine load for fixed and controllable pitch propeller drivers. Whenever an engine is driving a fixed pitch propeller, the Project Guide considers brake power and speed linked by the propeller law [29], whilst for controllable pitch propeller, speed is considered constant. Since these aspects were not considered, the main contribution of the present work is the development of an engine model suitable to optimize the selection of dual-fuel two-stroke Diesel engines considering the uniqueness of each engine and the propeller type to be driven.

## 2. Methodology

The algorithms included in the proposed model were implemented in a MatLab environment. Including engine data from various manufacturers would be ideal in this sort of study, but the required information is not readily provided by the most engine manufacturers. Therefore, owing to the broad data availability of a web-based application from a two-stroke Diesel engine manufacturer [28], only Diesel engines covered by this application were studied. Standard ambient conditions provided by the International Organization for Standardization and a sulphur content of 0.5% were assumed. Although engine type designation refers to the number of cylinders, stroke/bore ratio, diameter of piston, engine concept, mark number, fuel injection concept, and Tier III technology, narrow configurations of engines were studied. All the addressed engines were not equipped with Tier III technology, and they held the same fuel injection concept (GI) and engine concept (ME-C). Furthermore, only default configurations of engines were taken into account.

Two-stroke marine engine selection begins with placing the SMCR point on the engine layout diagram programme to identify engines able to supply the required power and speed. An engine layout diagram is an envelope that defines the field where nominal maximum firing pressure is available [30]. Every single engine holds a layout diagram depending on its number of cylinders, such that necessary information to plot the layout diagrams for the 22 engines considered herein is presented in Table 1. In this table, the brake power per cylinder on the four points of the envelope is listed ($P_{Bc,L1}$, $P_{Bc,L2}$, $P_{Bc,L3}$ and $P_{Bc,L4}$), as well as speed limits ($n_{min}$ and $n_{max}$) and limitations on the number of cylinders ($Z_{c,min}$ and $Z_{c,max}$) for every single engine. As it may be noticed, only engines of type G (green ultra-long stroke) and S (super long stroke), with diameters from 40 to 95 cm and various mark numbers (8.5, 9.5, 9.6...) were studied.

Since specific fuel consumption at SMCR depends on its position on the engine layout diagram, the SMCR was placed on the points L1, L2, L3, and L4, and the specific fuel consumptions of every engine were analyzed. However, specific fuel consumptions at part load depend also on the driven propeller type: fixed pitch propeller (FPP) or controllable pitch propeller (CPP). Hence, the web-based application was run four times for each of the 22 engines, considering FPP and CPP driving, summing up 176 runs. The application provides a table with specific fuel consumption [g/kWh] with loads from 10 to 100% of SMCR. Thus, specific fuel consumptions were normalized and the arising trends were approximated by polynomials.

**Table 1.** Available ME-GI slow-speed dual-fuel engines and their particulars to chart layout diagrams.

| Engine | $P_{Bc,L1}$ | $P_{Bc,L2}$ | $P_{Bc,L3}$ | $P_{Bc,L4}$ | $n_{min}$ | $n_{max}$ | $Z_{c,min}$ | $Z_{c,max}$ |
|---|---|---|---|---|---|---|---|---|
| | kW/Cylinder | | | | rpm | | Cylinder | |
| G95ME-C9.6 | 6870 | 5170 | 6440 | 4840 | 75 | 80 | 5 | 12 |
| G95ME-C10.5 | 6870 | 5170 | 6010 | 4520 | 70 | 80 | 5 | 12 |
| G95ME-C9.5 | 6870 | 5170 | 6010 | 4520 | 70 | 80 | 5 | 12 |
| G90ME-C10.5 | 6240 | 4670 | 5350 | 4010 | 72 | 84 | 5 | 12 |
| S90ME-C10.5 | 6100 | 4880 | 5230 | 4180 | 72 | 84 | 5 | 12 |
| G80ME-C9.5 | 4710 | 3550 | 3800 | 2860 | 58 | 72 | 6 | 9 |
| S80ME-C9.5 | 4510 | 3610 | 4160 | 3330 | 72 | 78 | 6 | 9 |
| G70ME-C10.5 | 3060 | 2540 | 2620 | 2180 | 66 | 77 | 5 | 6 |
| G70ME-C9.5 | 3640 | 2740 | 2720 | 2050 | 62 | 83 | 5 | 8 |
| S70ME-C10.5 | 3430 | 2580 | 2750 | 2070 | 73 | 91 | 5 | 8 |
| S70ME-C8.5 | 3270 | 2610 | 2620 | 2100 | 73 | 91 | 5 | 8 |
| S65ME-C8.5 | 2870 | 2290 | 2330 | 1860 | 77 | 95 | 5 | 8 |
| G60ME-C9.5 | 2680 | 2010 | 1990 | 1500 | 72 | 97 | 5 | 8 |
| S60ME-C10.5 | 2490 | 1880 | 2000 | 1500 | 84 | 105 | 5 | 8 |
| S60ME-C8.5 | 2380 | 1900 | 1900 | 1520 | 84 | 105 | 5 | 8 |
| G50ME-C9.6 | 1720 | 1290 | 1360 | 1020 | 79 | 100 | 5 | 9 |
| S50ME-C9.7 | 1780 | 1340 | 1290 | 970 | 85 | 117 | 5 | 9 |
| S50ME-C9.6 | 1780 | 1420 | 1350 | 1080 | 89 | 117 | 5 | 9 |
| S50ME-C8.5 | 1660 | 1330 | 1340 | 1070 | 102 | 127 | 5 | 9 |
| G45ME-C9.5 | 1390 | 1045 | 1090 | 820 | 87 | 111 | 5 | 8 |
| G40ME-C9.5 | 1100 | 825 | 870 | 655 | 99 | 125 | 5 | 8 |
| S40ME-C9.5 | 1135 | 910 | 865 | 690 | 111 | 146 | 5 | 9 |

## 2.1. Specific Fuel Consumption at SMCR

Firstly, specific fuel consumptions at SMCR were divided by themselves at the nominal maximum continuous rating (NMCR) to obtain the normalized specific fuel consumptions regarding NMCR: $SFOC_N$, $SGC_N$ and $SPOC_N$. Equation (1) mathematically describes this procedure for each one of the normalized specific fuel consumption ($SFC_N$). In this equation, $P_B$ is brake power [kW]; $j$ varies from 1 to 4, representing the SMCR position (L1, L2, L3, and L4); and $k$ varies from 1 to 22, representing the engines. Then, regressions were performed as a function of mean effective pressure and engine speed, normalized with respect to NMCR ($MEP_N$ and $n_N$), as respectively defined by Equations (2) and (3). Mean effective pressure may also be written as in Equation (4) [29]. Therefore, knowing that the number of cylinders ($Z_c$), revolutions of crankshaft per complete working cycle ($r$), and cylinder swept volume ($V_S$) are engine constants, $MEP_N$ could also be written as in Equation (5). Hence, $n_N$ and $MEP_N$ could be calculated with the support of Table 1.

$$SFC_{N,jk} = \frac{SFC_{SMCR,jk}}{SFC_{NMCR,k}} \quad (1)$$

$$MEP_{N,jk} = \frac{MEP_{SMCR,jk}}{MEP_{NMCR,k}} \quad (2)$$

$$n_{N,jk} = \frac{n_{SMCR,jk}}{n_{NMCR,k}} \quad (3)$$

$$MEP = \frac{r}{Z_c \cdot V_S} \cdot \frac{P_B}{n} \quad (4)$$

$$MEP_{N,jk} = \frac{P_{SMCR,jk}}{n_{SMCR,jk}} \cdot \frac{n_{NMCR,k}}{P_{NMCR,k}} \quad (5)$$

The polynomial surfaces obtained for specific fuel consumptions at SMCR normalized with respect to NMCR and their percentage errors of regression are illustrated in Figure 1. As it can be seen, normalized specific fuel consumptions vary almost linearly with respect to $MEP_N$ and are practically not influenced by $n_N$, such that they could be approached by plans. Moreover, engines of type G and S did not present substantial differences from each other and, for this reason, they were analyzed together. The largest deviations were 1.3%, 1.9%, and 1.4% for $SFOC_N$, $SPOC_N$, and $SGC_N$, respectively, showing that the polynomials achieved were suitable. In order to reproduce the polynomial surfaces, Table 2 provides the coefficients ($a$) for each one of the normalized specific fuel consumptions formulated as in Equation (6).

$$SFC_N = a_{00} + a_{10} \cdot n_N + a_{01} \cdot MEP_N \tag{6}$$

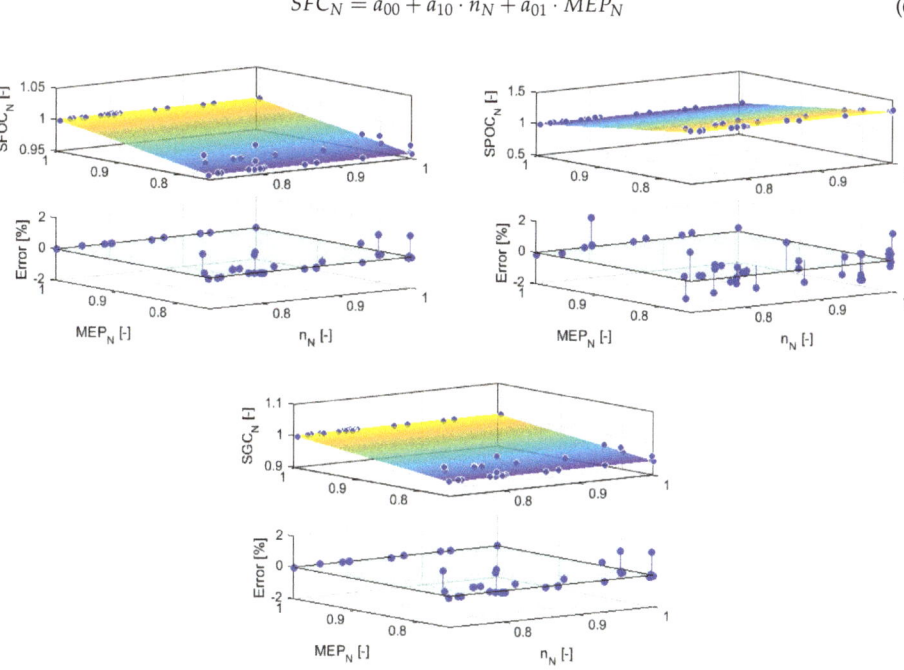

**Figure 1.** Polynomial surfaces of the specific fuel consumptions normalized with respect to the nominal maximum continuous rating (NMCR) and respective percentage errors of regression.

**Table 2.** Coefficients of the polynomial surfaces.

| Coefficients | $SFOC_N$ | $SPOC_N$ | $SGC_N$ |
|---|---|---|---|
| $a_{00}$ | 0.8342 | 2.294 | 0.7870 |
| $a_{10}$ | −0.0009009 | −0.0006027 | −0.000691 |
| $a_{01}$ | 0.1665 | −1.296 | 0.2136 |

## 2.2. Specific Fuel Consumption at Part Load

The normalized specific fuel consumptions with respect to SMCR ($SFOC_S$, $SGC_S$, $SPOC_S$) were achieved by dividing their values at part load by themselves at SMCR. Equation (7) exemplifies this procedure for a generic normalized specific fuel consumption ($SFC_S$), where the index $i$ varies from 1 to 19, representing engine loads from 10 to 100% with 5% step; $j$ varies between 1 and 4 representing the SMCR position; and $k$ varies between 1 and 22, representing the engines.

$$SFC_{S,ijk} = \frac{SFC_{ijk}}{SFC_{SMCR,jk}} \qquad (7)$$

A compromise between simplicity and accuracy was pursued to develop really fast and useful modelling. Thus, polynomials with degrees from one to nine were tested, and the one presenting the best fitting was selected. In some cases, the data set presented discontinuities too sharp to be captured by only one polynomial curve, even of a high order. Instead, two and three polynomials of a low order presented better fittings and were preferred in such cases.

Figures 2 and 3 show the normalized curves and their percentage errors of regression as a function of brake power given in a percentage of SMCR ($P_B[\%SMCR]$) for FPP and CPP driving, respectively. The specific fuel consumptions hold distinct behaviours from one another and, consequently, they were approximated by polynomials of different degrees. Furthermore, for a better approximation of the $SGC_S$ trend, three polynomials were employed. The driving type is not very influential on the form of the curves except on the $SGC_S$, which has slightly different behaviours for FPP and CPP driving. In general, the mismatches rise as engine load decreases, except regarding $SPOC_S$, whose errors are quite dispersed over the load range analyzed. Regarding FPP driving, the largest errors were 2.2%, −2.0%, and −2.5%, respectively for $SFOC_S$, $SPOC_S$, and $SGC_S$ (Figure 2). Similarly, the largest errors regarding CPP driving were 2.2%, 1.8%, and −2.1%, respectively for $SFOC_S$, $SPOC_S$, and $SGC_S$ (Figure 3).

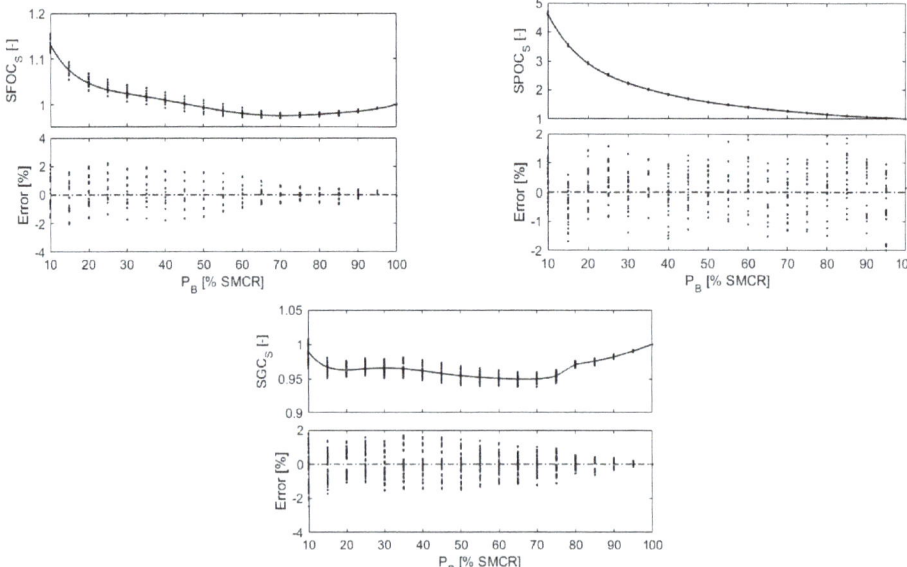

**Figure 2.** Polynomial curves of the specific fuel consumptions normalized with respect to specified maximum continuous rating (SMCR) and respective percentage errors of regression for fixed pitch propeller (FPP) driving.

All polynomial curves were obtained by using centring and scaling transformation to improve the numerical properties of both the polynomial and fitting algorithms. Table 3 provides the coefficients ($a$) for the formulation given in Equation (8), in which $x$ is a function of load [% SMCR], mean ($\mu$), and standard deviation ($\sigma$) of the applicable load range, as given in Equation (9). As the $SGC_S$ was approached by three polynomials, the letters "a", "b", and "c" in this table indicate the load range where each polynomial is applicable. Thus, these letters indicate, respectively, load ranges from 10 to 70%, 70 to 80%, and 80 to 100%.

$$SFC_S = a_0 + a_1 \cdot x + a_2 \cdot x^2 + a_3 \cdot x^3 + \ldots + a_7 \cdot x^7 \tag{8}$$

$$x = \frac{P_B[\%SMCR] - \mu[\%SMCR]}{\sigma[\%SMCR]} \tag{9}$$

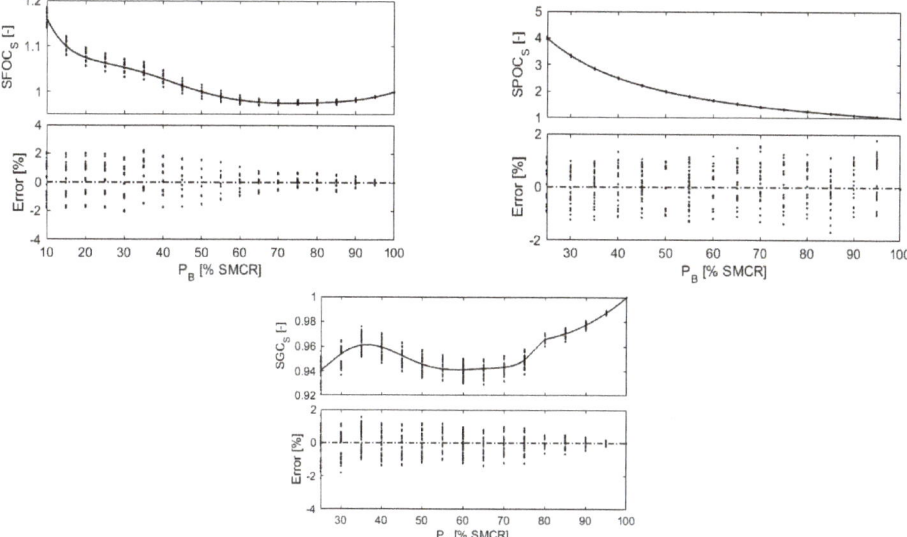

**Figure 3.** Polynomial curves of the specific fuel consumptions normalized with respect to SMCR and respective percentage errors of regression for controllable pitch propeller (CPP) driving.

**Table 3.** Coefficients of the polynomial curves.

| Driving | Coefficients | SFOC$_S$ | SPOC$_S$ | SGC$_S$ a | SGC$_S$ b | SGC$_S$ c |
|---|---|---|---|---|---|---|
| FPP | $a_0$ * | 986.3 | 1484 | 959.6 | 962.4 | 982.0 |
|  | $a_1$ * | −36.64 | −486.5 | −15.06 | 11.91 | 10.34 |
|  | $a_2$ * | 25.25 | 256.4 | −0.4842 | 0 | 1.730 |
|  | $a_3$ * | 20.94 | −134.4 | 12.56 | 0 | 0 |
|  | $a_4$ * | −13.29 | −47.19 | −5.405 | 0 | 0 |
|  | $a_5$ * | −8.209 | 36.87 | −4.242 | 0 | 0 |
|  | $a_6$ * | 5.533 | 49.79 | 2.848 | 0 | 0 |
|  | $a_7$ * | 0 | −26.58 | 0 | 0 | 0 |
|  | $\mu$ | 55.00 | 55.00 | 42.50 | 77.50 | 90.00 |
|  | $\sigma$ | 27.39 | 27.39 | 20.16 | 3.536 | 7.079 |
| CPP | $a_0$ * | 989 | 1599 | 945.6 | 957.7 | 977.5 |
|  | $a_1$ * | −49.71 | −596.8 | −17.73 | 11.95 | 11.99 |
|  | $a_2$ * | 50.68 | 221.8 | 16.44 | 0 | 2.789 |
|  | $a_3$ * | 4.908 | −62.2 | 9.05 | 0 | 0 |
|  | $a_4$ * | −31.54 | 20.48 | −13.56 | 0 | 0 |
|  | $a_5$ * | 7.997 | −22.79 | −0.3384 | 0 | 0 |
|  | $a_6$ * | 9.374 | 9.135 | 2.745 | 0 | 0 |
|  | $a_7$ * | −3.594 | 0 | 0 | 0 | 0 |
|  | $\mu$ | 55.00 | 55.00 | 42.50 | 77.50 | 90.00 |
|  | $\sigma$ | 27.39 | 27.39 | 20.16 | 3.536 | 7.079 |

* These coefficient's values are multiplied by 1000.

## 3. Model

The model relies on computing the specific fuel consumptions for any SMCR and any part load by means of Equation (10). Consequently, the polynomials previously given by Equations (6) and (8) and the coefficients listed in Table 2 must be used. To use these equations, it is also necessary to consider Table 1 to calculate MEP$_N$ and $n_N$, respectively by Equations (3) and (5) for a certain engine and SMCR. Additionally, the specific fuel consumptions at NMCR (SFOC$_{NMCR}$, SPOC$_{NMCR}$, and SGC$_{NMCR}$) for every single engine are necessary and, for this reason, they are listed in Table 4. The only difference between FPP and CPP driving is the coefficients' values of the polynomial SFC$_S$.

$$SFC = SFC_{NMCR} \cdot SFC_N \cdot SFC_S \tag{10}$$

**Table 4.** Specific fuel consumptions at NMCR [g/kWh].

| Engine | SFOC$_{NMCR}$ | SPOC$_{NMCR}$ | SGC$_{NMCR}$ |
|---|---|---|---|
| G95ME-C9.6 | 165.0 | 4.9 | 135.9 |
| G95ME-C10.5 | 163.0 | 4.9 | 134.2 |
| G95ME-C9.5 | 166.0 | 5.0 | 136.7 |
| G90ME-C10.5 | 165.0 | 4.9 | 135.9 |
| S90ME-C10.5 | 166.0 | 5.0 | 136.7 |
| G80ME-C9.5 | 166.0 | 5.0 | 136.7 |
| S80ME-C9.5 | 166.0 | 5.0 | 136.7 |
| G70ME-C10.5 | 163.0 | 5.5 | 133.7 |
| G70ME-C9.5 | 167.0 | 5.0 | 137.5 |
| S70ME-C10.5 | 166.0 | 5.0 | 136.7 |
| S70ME-C8.5 | 169.0 | 5.0 | 139.2 |
| S65ME-C8.5 | 169.0 | 5.0 | 139.2 |
| G60ME-C9.5 | 167.0 | 5.0 | 137.5 |
| S60ME-C10.5 | 166.0 | 5.0 | 136.7 |
| S60ME-C8.5 | 169.0 | 5.0 | 139.2 |
| G50ME-C9.6 | 167.0 | 5.0 | 137.5 |
| S50ME-C9.7 | 165.0 | 4.9 | 135.9 |
| S50ME-C9.6 | 167.0 | 5.0 | 137.5 |
| S50ME-C8.5 | 170.0 | 5.1 | 140.0 |
| G45ME-C9.5 | 170.0 | 5.1 | 140.0 |
| G40ME-C9.5 | 174.0 | 5.2 | 143.3 |
| S40ME-C9.5 | 172.0 | 5.1 | 141.7 |

## 4. Results and Discussion

In order to investigate the accuracy of the approach proposed in this paper, two engines of intermediary NMCR were simulated, and the results were compared with catalogue data [28]. As the polynomials were obtained considering SMCR on L1, L2, L3, and L4, it was necessary to investigate the approach accuracy for intermediate points. Therefore, the SMCR was additionally placed on the center of engine layout diagrams (LC), such that the engine 9G80ME-C9.5-GI was examined for an SMCR of 33,571 kW and 65 rpm, and the engine 5S50ME-C9.6-GI was examined for an SMCR of 7037 kW and 103 rpm.

Figure 4 illustrates the qualitative and quantitative performance of the developed approach employed on the engine 9G80ME-C9.5-GI driving a fixed pitch propeller. It is noticeable in Figure 4a that the polynomials are able to predict the behaviour of SFOC, SPOC, and SGC with only minor mismatches, even when SMCR is in the center of the layout diagram (LC). Although the SMCR was not placed on LC in the regression process, the biggest deviations did not occur for LC, except regarding specific pilot oil consumption ($SPOC_e$), achieving $-3.6\%$ at 95% load, as it can be seen in Figure 4b. The percentage error for the specific fuel oil consumption ($SFOC_e$) peaked at $-2.0\%$ when SMCR was on L3, and the load was 15% of SMCR. Also in a low load, the percentage error for the specific gas consumption ($SGC_e$) peaked at $-1.7\%$ when SMCR was on L2 and the load was 10% of SMCR.

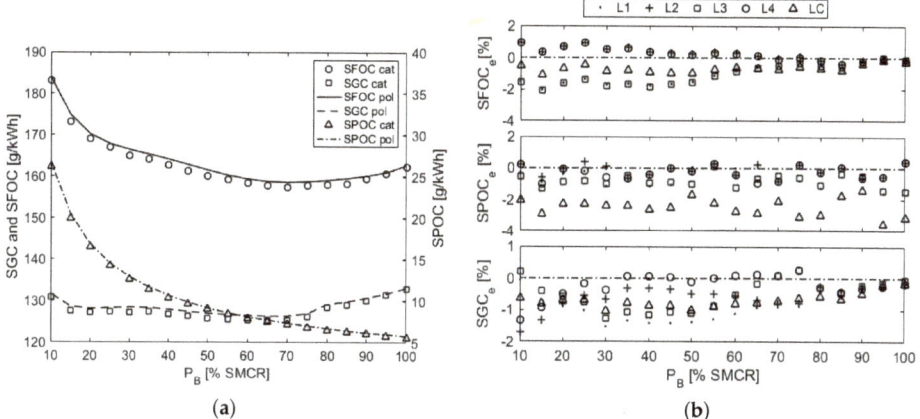

**Figure 4.** Qualitative and quantitative performance of the developed approach for the engine 9G80ME-C9.5-GI driving a fixed pitch propeller. (a) Qualitative comparison for the SMCR in the center of the layout diagram (LC). (b) Quantitative comparison for the SMCR in five different positions.

Qualitative and quantitative performances of the developed approach employed on the engine 5S50ME-C9.6-GI driving a controllable pitch propeller is illustrated in Figure 5. As in Figure 4a, it is noted by Figure 5a that the polynomials are able to predict the behaviour of specific fuel consumptions for SMCR in the center of the layout diagram (LC) also for this engine and CPP driving. Likewise, the biggest deviations did not occur for LC, except regarding $SPOC_e$, achieving $-2.8\%$ at 100% load, as it can be seen in Figure 5b. The percentage error $SFOC_e$ peaked at $-2.0\%$ when SMCR was on L1 and the load was 30% of SMCR. Either for 25% or 65% of SMCR, $SGC_e$ achieved $-1.6\%$ when SMCR was on L2.

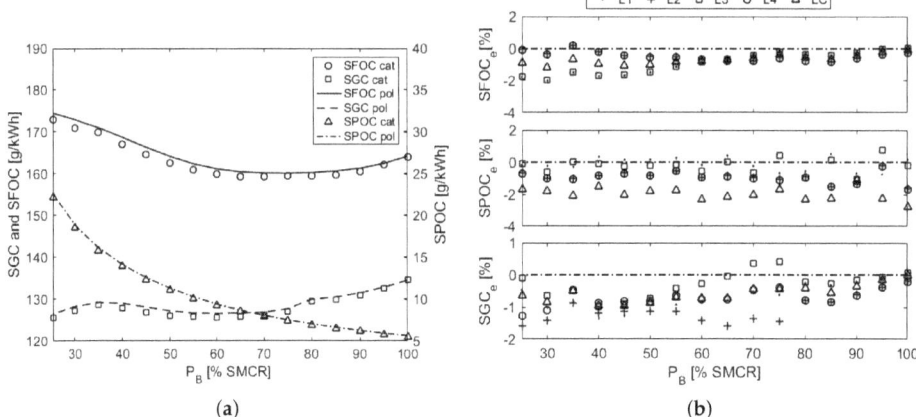

**Figure 5.** Qualitative and quantitative performance of the developed approach for the engine 5S50ME-C9.6-GI driving a controllable pitch propeller. (**a**) Qualitative comparison for the SMCR in the center of the layout diagram (LC). (**b**) Quantitative comparison for the SMCR in five different positions.

## 5. Conclusions

The present study has provided state-of-the-art modelling of marine engines and ship energy systems, and also addressed programming languages and dedicated applications to be employed in these areas. Moreover, a simple and fast model to assist in tackling optimization problems involving the selection of dual-fuel two-stroke Diesel engines was developed. The model was implemented in a MatLab environment and relies on normalizing specific fuel consumptions and approximating their trends by polynomials. Since the behaviour of marine engines driving fixed and controllable pitch propellers has been taken into account, different polynomials were derived.

Finally, the comparison between the model predictions and catalogue data for two engines of different sizes driving different propeller types revealed that the model was capable of adequately representing the behaviour of the specific fuel consumptions. The majority of the deviations was negative, meaning that the model overestimated the specific fuel consumptions of those engines. The greatest deviation did not exceed −3.6%, even when the specified maximum continuous rating was placed in the center of the layout diagram. Having this scenario was quite acceptable, and the model may be used successfully whenever one is interested in a fast and easy way to obtain specific fuel consumptions of several engines, such as in optimization problems.

**Author Contributions:** Conceptualization, all authors; Methodology, C.H.M. and J.-D.C.; Software, C.H.M.; Validation, C.H.M., J.-D.C. and C.R.P.B.; Formal Analysis, C.R.P.B. and A.M.; Investigation, C.H.M.; Resources, C.H.M.; Data Curation, C.H.M.; Writing—Original Draft Preparation, C.H.M.; Writing—Review & Editing, C.H.M.; Visualization, A.M.; Supervision, J.-D.C. and C.R.P.B.

**Funding:** This research received no external funding.

**Conflicts of Interest:** The authors declare no conflict of interest.

## Abbreviations

The following abbreviations are used in this manuscript:

| | |
|---|---|
| 0-D | zero-dimensional |
| CFD | computational fluid dynamics |
| CPP | controllable pitch propeller |
| FPP | fixed pitch propeller |
| MEP | mean effective pressure |
| MVEM | mean value engine model |
| NMCR | nominal maximum continuous rating |
| SFC | specific fuel consumption (generic) |
| SFOC | specific fuel oil consumption |
| SGC | specific gas consumption |
| SMCR | specified maximum continuous rating |
| SPOC | specific pilot oil consumption |
| TFEM | transfer function engine model |

## References

1. Thomson, H.; Corbett, J.J.; Winebrake, J.J. Natural gas as a marine fuel. *Energy Policy* **2015**, *87*, 153–167. doi:10.1016/j.enpol.2015.08.027. [CrossRef]
2. IMO. Resolution MEPC.176(58): Amendments to the Annex of the Protocol of 1997 to Amend the International Convention for the Prevention of Pollution from Ships, 1973, as Modified by the Protocol of 1978 Relating Thereto. 2008. Available online: http://www.imo.org/en/KnowledgeCentre/IndexofIMOResolutions/Marine-Environment-Protection-Committee-%28MEPC%29/Documents/MEPC.176%2858%29.pdf (accessed on 3 October 2018).
3. IMO. IMO Sets 2020 Date for Ships to Comply with Low Sulphur Fuel Oil Requirement. 2016. Available online: http://www.imo.org/en/MediaCentre/PressBriefings/Pages/MEPC-70-2020sulphur.aspx (accessed on 3 October 2018).
4. DNV-GL. Sulphur Cap Ahead—Time to Take Action. Available online: https://www.dnvgl.com/article/sulphur-cap-ahead-time-to-take-action-94198 (accessed on 3 October 2018).
5. Karim, G. *Dual-Fuel Diesel Engines*; CRC Press: Boca Raton, FL, USA, 2015. doi:10.1201/b18163.
6. Caton, J.A. (Ed.) *An Introduction to Thermodynamic Cycle Simulations for Internal Combustion Engines*; John Wiley & Sons, Ltd.: Hoboken, NJ, USA, 2015. doi:10.1002/9781119037576.
7. Schulten, P.J.M. The Interaction Between Diesel Engines, Ship and Propellers During Manoeuvring. Ph.D. Thesis, Delft University of Technology, Delft, The Netherlands, 2005.
8. Hountalas, D.T. Prediction of marine diesel engine performance under fault conditions. *Appl. Therm. Eng.* **2000**, *20*, 1753–1783. doi:10.1016/S1359-4311(00)00006-5. [CrossRef]
9. Rakopoulos, C.; Dimaratos, A.; Giakoumis, E.; Rakopoulos, D. Evaluation of the effect of engine, load and turbocharger parameters on transient emissions of diesel engine. *Energy Convers. Manag.* **2009**, *50*, 2381–2393. doi:10.1016/j.enconman.2009.05.022. [CrossRef]
10. Scappin, F.; Stefansson, S.H.; Haglind, F.; Andreasen, A.; Larsen, U. Validation of a zero-dimensional model for prediction of NOx and engine performance for electronically controlled marine two-stroke diesel engines. *Appl. Therm. Eng.* **2012**, *37*, 344–352. doi:10.1016/j.applthermaleng.2011.11.047. [CrossRef]
11. Cordtz, R.; Schramm, J.; Andreasen, A.; Eskildsen, S.S.; Mayer, S. Modeling the Distribution of Sulfur Compounds in a Large Two Stroke Diesel Engine. *Energy Fuels* **2013**, *27*, 1652–1660. doi:10.1021/ef301793a. [CrossRef]
12. Raptotasios, S.I.; Sakellaridis, N.F.; Papagiannakis, R.G.; Hountalas, D.T. Application of a multi-zone combustion model to investigate the NOx reduction potential of two-stroke marine diesel engines using EGR. *Appl. Energy* **2015**, *157*, 814–823. doi:10.1016/j.apenergy.2014.12.041. [CrossRef]
13. Payri, F.; Olmeda, P.; Martín, J.; García, A. A complete 0D thermodynamic predictive model for direct injection diesel engines. *Appl. Energy* **2011**, *88*, 4632–4641. doi:10.1016/j.apenergy.2011.06.005. [CrossRef]

14. Cordtz, R.; Mayer, S.; Eskildsen, S.S.; Schramm, J. Modeling the condensation of sulfuric acid and water on the cylinder liner of a large two-stroke marine diesel engine. *J. Mar. Sci. Technol.* **2017**, *23*, 178–187. doi:10.1007/s00773-017-0455-9. [CrossRef]
15. Gugulothu, S.; Reddy, K. CFD simulation of in-cylinder flow on different piston bowl geometries in a DI diesel engine. *J. Appl. Fluid Mech.* **2016**, *9*, 1147–1155. doi:10.18869/acadpub.jafm.68.228.24397. [CrossRef]
16. Pang, K.M.; Karvounis, N.; Walther, J.H.; Schramm, J. Numerical investigation of soot formation and oxidation processes under large two-stroke marine diesel engine-like conditions using integrated CFD-chemical kinetics. *Appl. Energy* **2016**, *169*, 874–887. doi:10.1016/j.apenergy.2016.02.081. [CrossRef]
17. Sun, X.; Liang, X.; Shu, G.; Lin, J.; Wang, Y.; Wang, Y. Numerical investigation of two-stroke marine diesel engine emissions using exhaust gas recirculation at different injection time. *Ocean Eng.* **2017**, *144*, 90–97. doi:10.1016/j.oceaneng.2017.08.044. [CrossRef]
18. Benvenuto, G.; Carrera, G.; Rizzuto, E. Dynamic Simulation of Marine Propulsion Plants. In Proceedings of the International Conference on Ship and Marine Research, Rome, Italy, 5–7 October 1994.
19. Kyrtatos, N.P.; Theodossopoulos, P.; Theotokatos, G.; Xiros, N. Simulation of the overall ship propulsion plant for performance prediction and control. In Proceedings of the Conference on Advanced Marine Machinery Systems with Low Pollution and High Efficiency, Newcastle upon Tyne, UK, 25–26 March 1999.
20. Theotokatos, G.P. Ship Propulsion Plant Transient Response Investigation using a Mean Value Engine Model. *Int. J. Energy* **2008**, *2*, 66–74.
21. Stapersma, D.; Vrijdag, A. Linearisation of a ship propulsion system model. *Ocean Eng.* **2017**, *142*, 441–457. doi:10.1016/j.oceaneng.2017.07.014. [CrossRef]
22. Michalski, J. A method for selection of parameters of ship propulsion system fitted with compromise screw propeller. *Pol. Marit. Res.* **2007**, *14*. doi:10.2478/v10012-007-0032-y. [CrossRef]
23. Aldous, L.; Smith, T. Speed Optimisation for Liquefied Natural Gas Carriers: A Techno-Economic Model. In Proceedings of the International Conference on Technologies, Operations, Logistics & Modelling for Low Carbon Shipping, Newcastle, UK, 11–12 September 2012.
24. Talluri, L.; Nalianda, D.; Kyprianidis, K.; Nikolaidis, T.; Pilidis, P. Techno economic and environmental assessment of wind assisted marine propulsion systems. *Ocean Eng.* **2016**, *121*, 301–311. doi:10.1016/j.oceaneng.2016.05.047. [CrossRef]
25. Sakalis, G.N.; Frangopoulos, C.A. Intertemporal optimization of synthesis, design and operation of integrated energy systems of ships: General method and application on a system with Diesel main engines. *Appl. Energy* **2018**, *226*, 991–1008. doi:10.1016/j.apenergy.2018.06.061. [CrossRef]
26. Trivyza, N.L.; Rentizelas, A.; Theotokatos, G. A novel multi-objective decision support method for ship energy systems synthesis to enhance sustainability. *Energy Convers. Manag.* **2018**, *168*, 128–149. doi:10.1016/j.enconman.2018.04.020. [CrossRef]
27. Shi, W.; Grimmelius, H.T.; Stapersma, D. Analysis of ship propulsion system behaviour and the impact on fuel consumption. *Int. Shipbuild. Prog.* **2010**, *57*, 35–64. doi:10.3233/ISP-2010-0062. [CrossRef]
28. MAN Diesel & Turbo. *Computerised Engine Application System-Engine Room Dimensioning (CEAS-ERD)*; MAN Diesel & Turbo: Munich, Germany, 2018.
29. Woud, H.K.; Stapersma, D. *Design of Propulsion and Electric Power Generation Systems*; IMarEST: London, UK, 2013.
30. Woodyard, D. *Pounder's Marine Diesel Engines and Gas Turbines*; Elsevier BV: Oxford, UK, 2009. doi:10.1016/b978-0-7506-8984-7.00002-3.

© 2019 by the authors. Licensee MDPI, Basel, Switzerland. This article is an open access article distributed under the terms and conditions of the Creative Commons Attribution (CC BY) license (http://creativecommons.org/licenses/by/4.0/).

*Article*

# Numerical Study of Turbulent Air and Water Flows in a Nozzle Based on the Coanda Effect

Youssef El Halal [1], Crístofer H. Marques [1], Luiz A. O. Rocha [2], Liércio A. Isoldi [1], Rafael de L. Lemos [1], Cristiano Fragassa [3] and Elizaldo D. dos Santos [1,*]

1. School of Engineering, Universidade Federal do Rio Grande–FURG, Rio Grande, RS 96203-900, Brazil; youssefhalal20@gmail.com (Y.E.H.); cristoferhood@gmail.com (C.H.M.); liercioisoldi@furg.br (L.A.I.); er.lemos@outlook.com (R.d.L.L.)
2. Programa de Pós-Graduação em Engenharia Mecânica, Universidade do Vale do Rio dos Sinos (UNISINOS), São Leopoldo, RS 93022-750, Brazil; luizor@unisinos.br
3. Department of Industrial Engineering, Alma Mater Studiorum Università di Bologna, Viale del Risorgimento 2, 40136 Bologna, Italy; cristiano.fragassa@unibo.it
* Correspondence: elizaldosantos@furg.br; Tel.: +55-53-3233.6916

Received: 16 November 2018; Accepted: 12 January 2019; Published: 22 January 2019

**Abstract:** In the present work it is performed a numerical study for simulation of turbulent air and water flows in a nozzle based on the Coanda effect named H.O.M.E.R. (High-Speed Orienting Momentum with Enhanced Reversibility). The main purposes of this work are the development of a numerical model for simulation of the main operational principle of the H.O.M.E.R. nozzle, verify the occurrence of the physical principle in a device using water as working fluid and generate theoretical recommendations about the influence of the difference of mass flow rate in two inlets and length of septum over the fluid dynamic behavior of water flow. The time-averaged conservation equations of mass and momentum are solved with the Finite Volume Method (FVM) and turbulence closure is tackled with the $k$-$\varepsilon$ model. Results for air flow show a good agreement with previous predictions in the literature. Moreover, it is also noticed that this main operational principle is promising for future applications in maneuverability and propulsion systems in marine applications. Results obtained here also show that water jets present higher deflection angles when compared with air jets, enhancing the capability of impose forces to achieve better maneuverability. Moreover, results indicated that the imposition of different mass flow rates in both inlets of the device, as well as central septum insertion have a strong influence over deflection angle of turbulent jet flow and velocity fields, indicating that these parameters can be important for maneuverability in marine applications.

**Keywords:** coanda effect; turbulence model; computational fluid dynamic; finite volume method; H.O.M.E.R. nozzle

## 1. Introduction

In 1936, Henri Coanda patented the first device capable to deflect a stream without any movable parts [1]. This phenomenon is named Coanda effect and consists on the tendency of a fluid to adhere to a curved surface due to the local pressure drop caused by acceleration of flow around a solid surface. Coanda surfaces are generated by asymmetric profiles configured in such as way that the flow properties in the exit of devices are changed [2]. One illustration of the domain subjected to Coanda effect flow can be seen in Figure 1. Thus, the Coanda effect became the target of many studies where several mechanisms which were based on this effect were developed, such as cooling [3–5], safety devices [6], aerospace applications [7–10] and few recent studies in marine devices [11,12]. The Coanda effect is also important for the design of Unmanned Aerial Vehicles (UAV). Some advantages on the use of this main operational principle for UAV as the capability to control the direction of the net force and improvement of lift forces acting on the device have been reported in the literature [13–17].

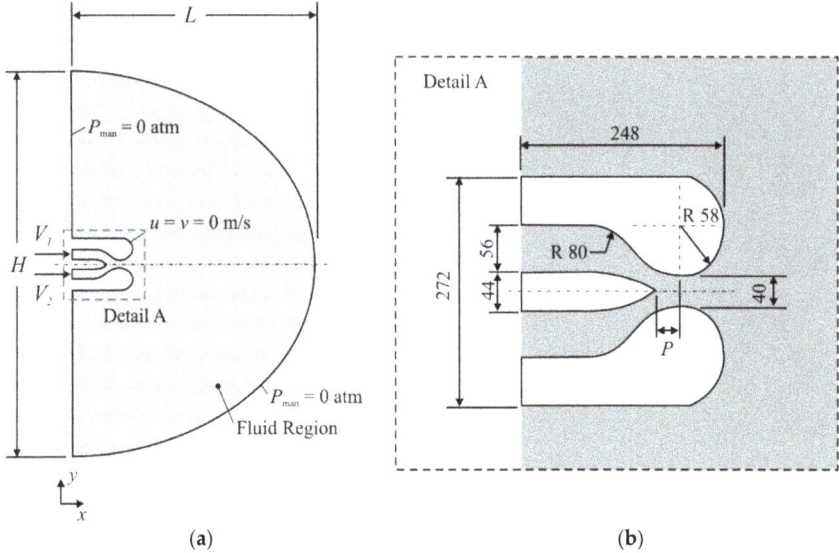

**Figure 1.** Computational domain of the studied problem: (**a**) entire domain and boundary conditions, (**b**) detailed view of the device (dimensions in mm).

Important works have been performed with the aim to improve the comprehension about the behavior of turbulent flows over aerodynamical profiles considering the Coanda effect. For instance, Ameri [18] conducted a theoretical and experimental study in ejectors with the aim to develop models for prediction of device performance based in this main operational principle. According to the author, in spite of large use of this principle in aeronautical applications, the challenge to understand the physical phenomenon leads to difficulties for design of the devices. Afterwards, Kim et al. [19] studied numerically the effect of some parameters, as pressure ratio between primary and secondary nozzles, over the performance of a Coanda nozzle. Djojodihardjo et al. [20] performed a numerical study using the Coanda effect applied to the control of wind turbines. More precisely, it is shown how the effect of a jet over the aerodynamical profile can be used to augment the lift and decrease the drag in the turbine blades. Gan et al. [21] simulated a two-dimensional jet flow over a logarithmic surface seeking to evaluate the behavior of the oncoming jet over the profile. Moreover, a parametric study about the effect of the outlet height of device over fluid dynamic behavior of jet flow was performed. Afterwards, Trancossi et al. [9] presented a work concerned with the application of a turbine based on a nozzle named ACHEON (Aerial Coanda High Efficiency Orienting Nozzle). The studied configuration allowed to perform a selective adhesion of jet flows (in two streams) over two Coanda surfaces. Moreover, some theoretical recommendations about parameters as flight autonomy was obtained, energy consumption and forces acting in profiles evaluating the possibility of application of this main operational principle for design of airplanes.

However, in the naval industry the study of Coanda devices is underestimated in spite of several upgrades that can be made to improve efficiency in this field. During a journey, for example, rudders angles of attack fluctuate between -10° and 10° [22], causing an augmentation of propulsion power necessary to keep the ship velocity due to drag forces on the rudder. Then, the employment of nozzles to maneuverability can be an important application to minimize this kind of problem. The H.O.M.E.R. (High-Speed Orienting Momentum with Enhanced Reversibility) nozzle technology has proven to be efficient for maneuverability of air vehicles [23]. Despite this new technology being proposed to aerospace vehicles, the main operational principle shall not be limited to air flows. In this context, the understanding of behavior of water flow in H.O.M.E.R nozzle is essential to verify its applicability for marine applications. A first attempt in this sense was performed in Lemos et al. [11], where a numerical study of turbulent water flows in two hydrodynamic profiles simulating the main operational principle of a hydro-propulsion device based on the Coanda effect was proposed. The influence of the distance between hydrodynamic profiles over mass flow rate, velocity and pressure fields were investigated and the results indicated a viability of this kind of main operational principle for water flows. Regardless of several important studies performed in the literature about turbulent flows in devices based on the Coanda effect, to the authors knowledge, only few studies considered water as working fluid, mainly in the H.O.M.E.R device. The present work is based on a proof of concept, i.e., if the main operational principle works for the present cases, it will have technical viability to be used for the design in marine applications for propulsion or maneuverability.

The present work aims to analyze numerically air and water flows in a H.O.M.E.R nozzle similar to that proposed by Trancossi and Dumas [24]. Both working fluids are considered incompressible. Moreover, turbulent flows at the steady state in a two-dimensional domain are simulated. The time-averaged conservation equations of mass and momentum are solved with the Finite Volume Method (FVM) [25–27]. To solve the closure problem of turbulence, the $k$-$\varepsilon$ model is employed [28,29]. Here, the numerical study is not concerned with the development of numerical methods for simulation of turbulent flows. However, important studies in the literature have been devoted for this kind of analysis [30,31]. In the present work, a first investigation of air flow in a H.O.M.E.R nozzle is simulated and results for deflection angle of jet are compared with those obtained by Trancossi and Dumas [24]. Afterwards, new theoretical recommendations about the fluid dynamic behavior of water flows in the device are obtained with the present numerical model. More precisely, the influence of the ratio between the mass flow rate in two inlets ($m^*$) over deflection angle of jet ($\alpha$) and velocity magnitudes of jet is evaluated. Moreover, the influence of $m^*$ on fluid dynamic behavior of flow in nozzle is investigated for two different septum insertions ($P = 10$ mm and 48 mm).

## 2. Methodology

### 2.1. Description of the Problem

The study is performed through several numerical simulations using the FVM, more precisely with the commercial CFD (Computational Fluid Dynamic) software package FLUENT 14.0 [25]. The computational domain is similar to that presented by Trancossi and Dumas [24]. Figure 1a shows the whole computational domain of the problem and its respective boundary conditions, while Figure 1b shows a detailed nozzle view with its dimensions. Once two different magnitudes of variable $P = 10$ mm and 48 mm (Figure 1b) are evaluated in the present work for air and water flows, this variable of septum insertion has no fixed magnitude in the sketch.

The H.O.M.E.R. nozzle is based on the Coanda effect. The main operational principle consists in forcing two fluid jets against two convex surfaces separated by a central septum (Figure 1a). Here, these two jets are imposed with velocities $V_1$ and $V_2$ in the upper and lower inlets, respectively. Those jets will adhere to the curved walls (where non-slip and impermeability boundary conditions are imposed) as a result of the differential pressure created by the viscous effects on the wall. This pressure gradient is mainly responsible for this attachment, once the ambient pressure is higher than the pressure near the wall. The end of the septum is characterized as a convergence zone of two imposed jets, starting the jets mixing process. The jet with the higher momentum will drag the other along, causing flow deflection. In the present simulations, the mixed jet flows in a region defined by a semi-circular area (gray area in Figure 1a) with dimensions $H = 2L = 1000$ mm. This domain is defined in such way that its external boundary conditions do not affect the flow near the nozzle. More precisely, atmospheric pressure in the exit lines of computational domain ($p_{man} = 0$ atm) is imposed.

Since the present work aims to analyze the behavior of flow in the nozzle using water as working fluid, a comparison of deflection angle of jet and velocity magnitudes in the $x$ and $y$ directions with air flows is made. In order to evaluate the flow deflection ($\alpha$), a total mass flow rate ($\dot{m}$) equals to 8.0 kg/s of water will be adopted. This magnitude is based on the study of Trancossi and Dumas [24] and represents values that can be found in real applications.

The deflection is a result of the difference between mass flow rate (which leads to difference of the momentum) of the flow in the superior and the inferior channels. This difference can be expressed in dimensionless form ($m^*$) by the following expression [24]:

$$m^* = \frac{\dot{m}_1 - \dot{m}_2}{\dot{m}_1 + \dot{m}_2} \quad (1)$$

here $\dot{m}_1$ and $\dot{m}_2$ are the mass flow rate in the superior and inferior channels, respectively. Those deflections will be measured with the software Digimizer [32], from the $x$ axis in anticlockwise direction to the central region of the jet. However, the sum of $\dot{m}_1$ and $\dot{m}_2$ will always be equal to $\dot{m}$.

In addition, the magnitude velocity ($V$) is measured on the nozzle outlet, located tangentially to the two curved surfaces. Then, the $x$ and $y$ velocities ($V_x$ and $V_y$, respectively) are calculated through the following equations:

$$V_x = V * \cos\alpha \quad (2)$$

$$V_y = V * \sin\alpha \quad (3)$$

## 2.2. Mathematical and Numerical Modeling

The study deals with a turbulent, incompressible and steady state flow in a two-dimensional domain. The standard $k$-$\varepsilon$ model (previously used in the study of Dumas and Trancossi [24]) was chosen to close the time-average equations. Considering this, the time-averaged conservation equations of mass and momentum are described by [29,33–35]:

$$\frac{\partial \overline{u}_i}{\partial x_i} = 0 \quad (4)$$

$$\frac{\partial}{\partial x_i}(\rho \overline{u}_i \overline{u}_j) = -\frac{\partial \overline{p}}{\partial x_j} + \frac{\partial}{\partial x_j}\left[\mu\left(\frac{\partial \overline{u}_i}{\partial x_i} + \frac{\partial \overline{u}_j}{\partial x_i}\right) - \rho \overline{u'_i u'_j}\right] \quad (5)$$

where $u_i$ is the velocity in the $i$–direction, with $i$ = 1 or 2 representing, respectively, the $x$ and $y$ direction. Moreover, $p$ represents the pressure, $\rho$ is the fluid density, $\mu$ is the fluid viscosity, the overbar is the time average operator and $'$ represents the fluctuation fields of pressure and velocity.

The Reynolds stress can be related to the time-averaged deformation rate by:

$$\overline{u'_i u'_j} = \frac{\mu_T}{\rho}\left(\frac{\partial \overline{u}_i}{\partial x_j} + \frac{\partial \overline{u}_j}{\partial x_i}\right) - \frac{2}{3}k\delta_{ij} \quad (6)$$

where $\delta_{ij}$ is the Kronecker delta and the eddy viscosity ($\mu_T$) can be described, for the $k$-$\varepsilon$ model, through the following equation [28]:

$$\mu_T = \rho C_\mu \frac{k^2}{\varepsilon} \quad (7)$$

where $C_\mu$ is a dimensionless constant (Table 1).

To obtain the turbulent kinetic energy ($k$) and its dissipation rate ($\varepsilon$) it is necessary to solve two additional transport equations given by:

$$\frac{\partial}{\partial x_i}(\rho u_i k) = \frac{\partial}{\partial x_i}\left[\left(\mu + \frac{\mu_t}{\sigma_k}\right)\frac{\partial k}{\partial x_i}\right] + G_k - \rho\varepsilon - Y_M + \sigma_K \quad (8)$$

$$\frac{\partial}{\partial x_i}(\rho u_i \varepsilon) = \frac{\partial}{\partial x_i}\left[\frac{\partial \varepsilon}{\partial x_i}\left(\mu + \frac{\mu_t}{\sigma_\varepsilon}\right)\right] + C_{1\varepsilon}\frac{\varepsilon}{k}G_k - C_{2\varepsilon}\rho\frac{\varepsilon^2}{k} + \sigma_\varepsilon \tag{9}$$

where $G_k$ is the turbulent kinetic rate and $Y_M$ is the contribution of the turbulence to the total dissipation rate. The constants $C_{1\varepsilon}$, $C_{2\varepsilon}$, $\sigma_k$ and $\sigma_\varepsilon$ are shown in Table 1 [28].

Concerning the closure modeling, it is worth to mention that several important studies in the literature [35–37] have recommended the employment of the $k$-$\omega$ or SST $k$-$\omega$ models for simulation of shear flows as those found in external flows over cylinders, bluff bodies and jets. In spite of this recommendation, in the present work the $k$-$\varepsilon$ model was used due to the fact that the results of Trancossi and Dumas [24] were achieved with this RANS closure model. Another important aspect is that the same cases simulated here with the $k$-$\varepsilon$ model were repeated with the SST $k$-$\omega$ model and results were similar (with differences lower than 0.1%). These results are not presented in the present paper.

Table 1. Constants employed in the numerical model.

| $C_\mu$ | $C_{1\varepsilon}$ | $C_{2\varepsilon}$ | $\sigma_\varepsilon$ | $\sigma_k$ |
|---|---|---|---|---|
| 0.09 | 1.44 | 1.92 | 1.3 | 1.0 |

To solve the time-averaged equations and the additional transport equations presented, the FVM is used, more precisely, the Ansys Fluent CFD code (version 14.0) [25]. The solver is pressure based [25,38]. To treat advective terms, the Second Order Upwind advection scheme is utilized and for the pressure-velocity coupling the SIMPLEC method is applied. Residuals of $10^{-6}$ are employed for the conservation equations of the mass and momentum and the transport equations of $k$ and $\varepsilon$.

Concerning the spatial discretization, an illustration of the independent grid employed here is shown in Figure 2a. The independent grid is composed of 40,000 finite triangular and rectangular volumes. It is worthy to mention that 4 different meshes with 10,000, 20,000, 40,000 and 80,000 were simulated. For the sake of brevity, results for mesh independence test are not presented here. Moreover, 20 layers of inflation are inserted near the device walls to capture the turbulent boundary layer, as well as to stabilize quantities (as drag coefficient) at every iteration. A detailed view is shown in Figure 2b. Due to the lack of precision of the $k$-$\varepsilon$ model to predict anisotropic gradients, the Enhanced Wall Treatment function was applied for the present simulations.

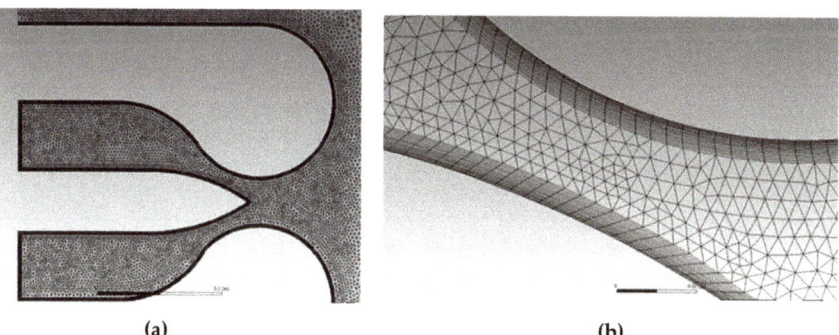

**Figure 2.** Spatial discretization of the problem. (**a**) Overview of the mesh. (**b**) Detail of the mesh in the high momentum zone.

List of symbols and abbreviations can found in Appendix A.

## 3. Results and Discussion

### 3.1. Computational Model Verification

In order to evaluate the present numerical model, which is one of the purposes here, a verification is performed adopting the model established by Trancossi and Dumas [24]. Six different flow cases for each working fluid with different $m^*$ are simulated and the deflection angles ($\alpha$) are compared with those found by the authors. The magnitudes of $m^*$ adopted here and the mass flow rates injected in the superior and inferior channels are presented in Table 2.

**Table 2.** The $m^*$ chosen and mass flow rates imposed in each channel of the device.

| $m^*$ | $\dot{m}_1$ (kg/s) | $\dot{m}_2$ (kg/s) |
|---|---|---|
| 0.00 | 4.0 | 4.0 |
| 0.10 | 4.4 | 3.6 |
| 0.20 | 4.8 | 3.2 |
| 0.25 | 5.0 | 3.0 |
| 0.50 | 6.0 | 2.0 |
| 0.75 | 7.0 | 1.0 |

Figure 3 shows the comparison between the jet deflected angles ($\alpha$) as a function of dimensionless difference of mass flow rate ($m^*$) obtained with the present model and those predicted by Trancossi and Dumas [24]. Then, it can be observed that the results obtained in the present simulations have the same tendency predicted in the literature. For $m^* > 0.2$ it can be observed that the present results underestimate the deflection angle in comparison with numerical predictions of Trancossi and Dumas [24]. One possible explanation for the differences found here can be related with uncertainties introduced by two equations turbulence models, as has been reported in detail in literature reviews [39–41]. According to the authors, the RANS-based models have inherent inability to replicate fundamental turbulence processes, which leads to difficulties for the prediction of many base flows, mainly external flows and jets. Some difficulties for numerical prediction of jets with RANS models are properly presented in Mishra and Iaccarino [41]. In spite of some differences found for the work of Ref. [24] and some difficulties of RANS models for simulation of jet flows, the model employed here can be considered verified for the achievement of new theoretical recommendations for turbulent flows in the nozzle.

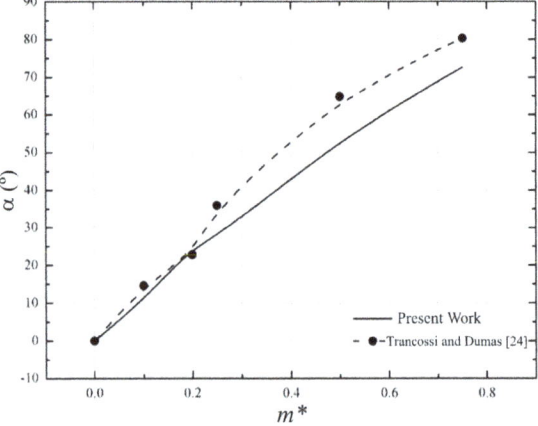

**Figure 3.** Comparison between the deflection angle of jet ($\alpha$) as a function of $m^*$ obtained in the present work and that predicted numerically by Trancossi and Dumas [24].

The next section is devoted to the simulation of water flows in the nozzle based on the Coanda effect considering two different septum insertion (P). More precisely, it is obtained the effect of $m^*$ over deflection angle ($\alpha$) and velocity magnitudes in $x$ and $y$ directions for water flows.

*3.2. Results for Water Flows in the H.O.M.E.R. Nozzle*

Initially, the influence of $m^*$ over the deflection angles ($\alpha$) for the two studied working fluids (air and water) is compared. Moreover, two different septum insertions were simulated ($P = 10$ mm and 48 mm) to investigate the influence of this variable on the deflection angle of air and water jets. Figure 4a,b show the results obtained injecting air and water in the device, respectively.

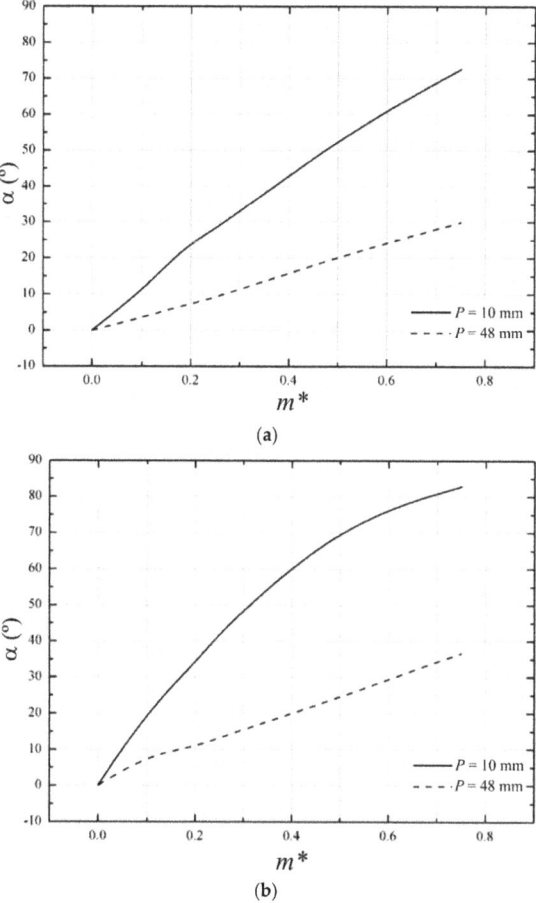

**Figure 4.** Deflection angles as function of $m^*$ for different working fluids and two different septum insertions P. (**a**) Air (**b**) Water.

Figure 4a shows that the parameter $P$ has a strong influence on the control of the deflection angle. For instance, an increase of nearly 2 times in magnitude of $\alpha$ for $P = 10$ mm in comparison with $P = 48$ mm is noticed. This increase is generated by a decrease of the channel height when the septum is inserted toward the convex walls. For $P = 10$ mm the fluid is forced to pass through a smaller space and it is accelerated. This increase in velocity leads to a decrease in absolute pressure, which raises the differential pressure between the flow and the ambient, and increases the space that the fluid remains attached to the surface. Furthermore, a similar behavior can also be observed when water is simulated as working fluid, Figure 4b. It is also noticed that the device delivers larger deflection angles when water is used, with a deflection angle nearly 12% higher than that reached for air as working fluid. This can be explained in terms of inviscid irrotational flow, as stated by Bradshaw [25]. Adopting the flow as inviscid, the differential pressure is proportional to the specific mass of the fluid. As water has higher density than air, it will remain attached longer. Another interesting behavior is observed when the effect of $m^*$ over $\alpha$ is predicted for $P = 10$ mm with air and water. For air flow, the increase of $\alpha$ with $m^*$ is almost linear, while for water the increase is more intense for lower magnitudes of $m^*$ and with the augmentation of $m^*$ the variation of $\alpha$ decreases. This difference can be associated with difference of viscosity of the fluids, which for the present configurations affects the Reynolds number of fluid flow in each inlet.

Since the H.O.M.E.R. nozzle is a device designed to improve maneuverability, it is important to evaluate the effect of parameters $m^*$ and $P$ on the outlet velocity, because the velocity has a dominant effect on fluid dynamic forces. Thus, Figure 5; Figure 6 show the velocities in the $y$ and $x$ direction, respectively, as a function of $m^*$ for two different insertions of the central septum ($P$).

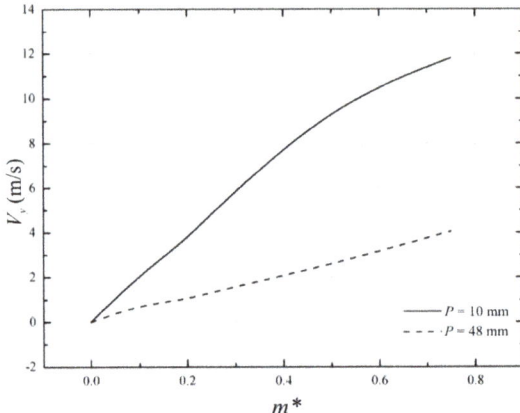

**Figure 5.** Effect of $m^*$ over magnitude of velocity in $y$ direction for two different values of $P$.

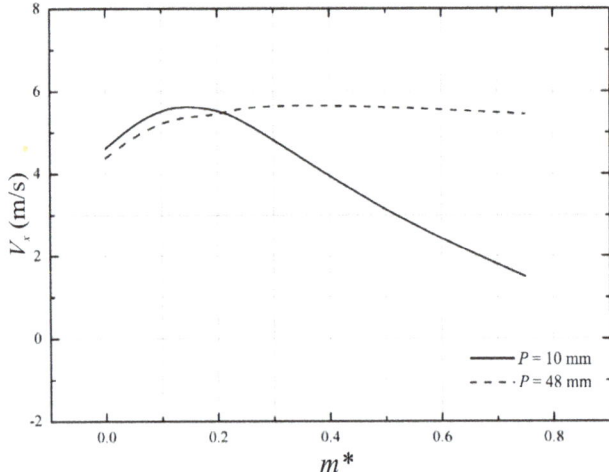

**Figure 6.** Effects of $m^*$ over magnitude of velocity in the $x$ direction for two different values of $P$.

Results indicated that the increase of $m^*$ also led to an increase of magnitude of velocity in the $y$ direction for both values of $P$, which is expected since the jet is deflecting towards the $y$ axis. Results also showed that the lower magnitude of $P$ led to a strong increase of $y$ direction velocity. Concerning the velocity in the $x$ direction, for $P = 10$ mm the opposite behavior is observed, i.e., the magnitude of the $x$ velocity decreases with the increase of $m^*$. For $P = 48$ mm the magnitude of the $x$ velocity is not much affected by variation of $m^*$. For lower magnitudes of $P$, results indicated that the strong restriction of channels increases the amount of momentum toward the $y$ direction, but suppresses the momentum in the $x$ direction. For the highest magnitude of $P$, the momentum magnitude is more equilibrated in both directions (with a tendency of increase for $y$ direction velocity). In general, results indicated that this kind of main operational principle seems suitable for maneuverability of marine applications. Moreover, the ratio between mass flow rates in two inlet channels and dimension of central septum insertion can be used to control the deflection angle of the mixture jet and its intensity (momentum).

For a better visualization of the flow, velocity magnitude fields can be observed in Figure 7. In this image the variation of deflection angle of jet between the insertion $P = 10$ mm (Figure 7a) and $P = 48$ mm (Figure 7b) is clear. Fields of Figure 7 also show the increase of jet deflection towards the $y$ axis with the increase of $m^*$. For $P = 10$ mm it can be seen that the jet is meagered when distant from the device as the magnitude of $m^*$ increases. For $P = 48$ mm, the magnitude of jet velocity is high along a longer distance from the device, but the jet is more concentrated in the central region of domain, i.e., with lower deflection angles in comparison with those observed for $P = 10$ mm.

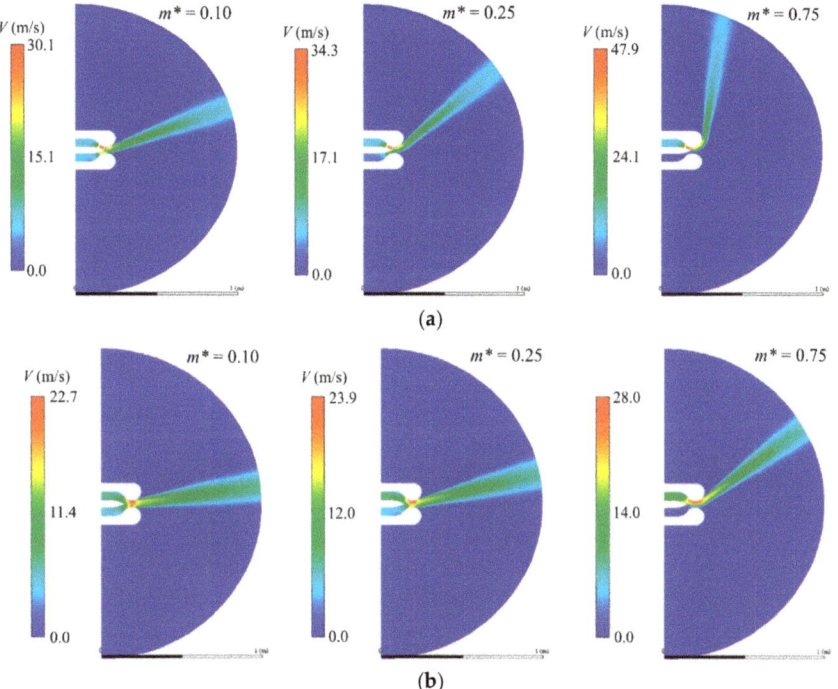

**Figure 7.** Velocity magnitude fields for different $m^*$: (**a**) $P = 10$ mm, (**b**) $P = 48$ mm.

As stated before, the differential pressure has a high influence on the detachment of a fluid jet on a curved surface. When the jet absolute pressure reaches the ambient pressure, the fluid will start to detach from surface. That phenomenon can be noticed in Figure 8, where the fluid start to change direction at the same moment that the jet pressure becomes equal to the external pressure.

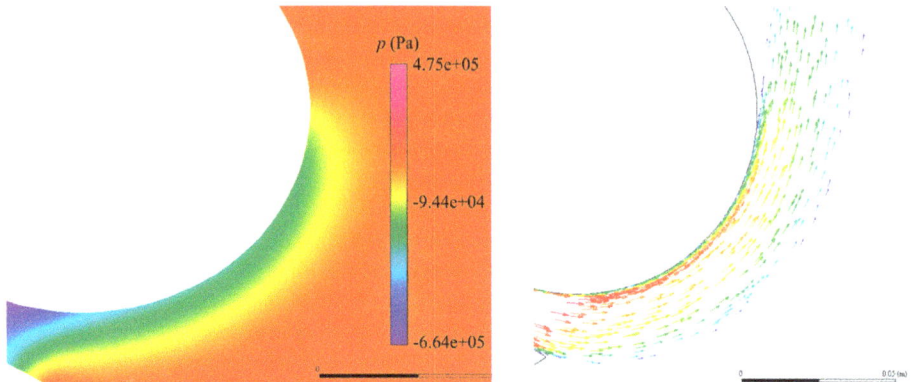

**Figure 8.** Absolute pressure field and vector velocity field for $P = 10$ mm and $m^* = 0.75$.

## 4. Conclusions

The present work showed promising results in relation to the use of water as working fluid in the H.O.M.E.R. nozzle. In general, the main operational principle based on the Coanda effect worked for water as working fluid. In this sense, the employment of this kind of main operational principle is technically feasible for the design of propulsion and/or maneuverability in marine applications. Results showed that the influence of the parameters $m^*$ and $P$ in the flow proved to be relevant, affecting the deflection angles magnitudes and velocities. When the insertion of the septum was equal to 10 mm, it shows angles 243% higher than the 48 mm insertion. However, the magnitude of velocity in the $x$ direction suffered a strong reduction for higher magnitudes of $m^*$, which indicates that the jet flow is more concentrated near the nozzle and a reduction of the driven force can occur in the device. In other words, the two studied parameters can be used to control the jet deflection and magnitude in the present problem.

For future works, the evaluation of other magnitudes of $P$ is recommended, as well as the study of forces acting in the device to obtain theoretical recommendations about the parameters that lead to the best performance of the device. Although the results proved to be favorable, an experimental model should be developed to validate the obtained numerical results.

**Authors Contributions**

Conceptualization, C.H.M., R.d.L.L. and E.D.S.; Formal analysis, C.H.M. and R.d.L.L.; Funding acquisition, L.A.O.R., L.A.I. and E.D.S.; Investigation, Y.E.H. and R.d.L.L.; Methodology, Y.E.H., C.H.M., L.A.O.R., L.A.I. and R.d.L.L.; Project administration, E.D.S.; Resources, C.F.; Software, Y.E.H. and R.d.L.L.; Supervision, E.D.S.; Validation, Y.E.H. and R.d.L.L.; Visualization, C.F.; Writing—original draft, Y.E.H. and E.D.S.; Writing—review & editing, C.H.M., L.A.O.R., L.A.I. and C.F.

**Funding:** FAPERGS, Research grants of CNPq (Processes: 306024/2017-9, 306012/2017-0, 307847/2015-2) and Universal Project (Process: 445095/2014-8).

**Acknowledgments:** The author Y.E. Halal thanks FAPERGS by Scientific Initiation Scholarship. The author R.L. Lemos thanks CNPq for a Master Science Scholarship. The authors E.D. Santos, L.A. Isoldi and L.A.O. Rocha thank CNPq (Brasília, DF, Brazil) for the research Grant (Processes: 306024/2017-9, 306012/2017-0, 307847/2015-2) and for financial support in the Universal project (Process: 445095/2014-8).

**Conflicts of Interest:** The authors declare no conflict of interest.

## Appendix A

*Appendix A.1. List of Symbols*

| | |
|---|---|
| $H$ | Height of domain (mm) |
| $k$ | Turbulent kinetic energy (m$^2$ s$^{-2}$) |
| $L$ | Length of domain (mm) |
| $m^*$ | Dimensionless mass flow rate |
| $\dot{m}_1$ | Mass flow rate on the superior channel (kg s$^{-1}$) |
| $\dot{m}_2$ | Mass flow rate on the inferior channel (kg s$^{-1}$) |
| $P$ | Septum insertion distance (mm) |
| $p$ | Pressure (Pa) |
| $u_i$ | Velocity in the $i$–direction (m s$^{-1}$) |
| $V_1$ | Velocity at the inlet of the superior channel (m s$^{-1}$) |
| $V_2$ | Velocity at the inlet of the inferior channel (m s$^{-1}$) |
| $x_i$ | Spatial coordinate in the $i$–direction (m) |
| $'$ | Fluctuation fields of pressure and velocity |
| — | Time average operator |
| $\alpha$ | Deflection angle of the mixed jet (°) |
| $\delta_{ij}$ | Kronecker delta |
| $\varepsilon$ | Dissipation rate (m$^2$ s$^{-3}$) |
| $\mu$ | Dynamic viscosity (kg m$^{-1}$ s$^{-1}$) |
| $\mu_T$ | Turbulent eddy viscosity (kg m$^{-1}$ s$^{-1}$) |
| $\rho$ | Density (kg m$^{-3}$) |

*Appendix A.2. List of Abbreviations*

| | |
|---|---|
| ACHEON | Aerial Coanda High Efficiency Orienting Nozzle |
| CFD | Computational Fluid Dynamics |
| FVM | Finite Volume Method |
| H.O.M.E.R. | High-Speed Orienting Momentum with Enhanced Reversibility |
| SIMPLEC | Semi-Implicit Method for Pressure Linked Equations-Consistent algorithm |

## References

1. Coanda, H. Device for Deflecting a Stream of Elastic Fluid Projected into an Elastic Fluid. U.S. Patent 2,052,869, 1 September 1936.
2. Constantin, O. Fluidic Elements based on Coanda Effect. *Incas Bull.* **2010**, *2*, 163–172. [CrossRef]
3. Schuh, H.; Persson, B. Heat transfer on circular cylinders exposed to free-jet flow. *Int. J. Heat Mass Transf.* **1964**, *7*, 1257–1271. [CrossRef]
4. Bhattacharya, A.; Wille, R. Der Einfluß von Nadeln auf die Kühlung von kurzen Kreiszylindern bei Freistrahlanblasung. *Glastechn. Ber.* **1959**, *32*, 397–401.
5. Sidiropoulos, V.; Vlachopoulos, J. An Investigation of Venturi and Coanda Effects in Blown Film Cooling. *Int. Polym. Process.* **2000**, *15*, 40–45. [CrossRef]
6. Stewart, B.A. Safety Device with Coanda Effect. U.S. Patent 3,909,037, 30 September 1975.
7. Wing, D.J. *Static Investigation of Two Fluidic Thrust-Vectoring Concepts on a Two-Dimensional Convergent-Divergent Nozzle*; NASA: Washington, DC, USA, 1994.
8. Mason, M.S.; Crowther, W.J. Fluidic thrust vectoring of low observable aircraft. In Proceedings of the CEAS Aerospace Aerodynamic Research Conference, Cambridge, UK, 10–12 June 2002.
9. Trancossi, M.; Madonia, M.; Dumas, A.; Angeli, D.; Bingham, C.; Das, S.S.; Grimaccia, F.; Marques, J.P.; Porreca, E.; Smith, T.; et al. A New Aircraft Architecture based on the ACHEON Coanda Effect Nozzle: Flight Model and Energy Evaluation. *Eur. Transp. Res. Rev.* **2016**, *8*, 11. [CrossRef]
10. Ahmed, R.I.; Abu Talib, A.R.; Mohd Rafie, A.S.; Djojodihardjo, H. Aerodynamics and flight mechanics of MAV based on Coanda Effect. *Aerosp. Sci. Technol.* **2017**, *62*, 136–147. [CrossRef]

11. Lemos, R.; Vieira, R.S.; Isoldi, L.A.; Rocha, L.A.O.; Pereira, M.S.; Dos Santos, E.D. Numerical analysis of a turbulent flow with Coanda effect in hydrodynamics profiles. *FME Trans.* **2017**, *45*, 412–420. [CrossRef]
12. Seo, D.-W.; Oh, J.; Jang, J. Performance Analysis of a Horn-Type Rudder Implementing the Coanda Effect. *Int. J. Nav. Archit. Ocean Eng.* **2017**, *9*, 177–184. [CrossRef]
13. Kulfan, B.M. Universal Parametric Geometry Representation Method. *J. Aircr.* **2008**, *45*, 142–158. [CrossRef]
14. Barlow, C.; Lewis, D.; Prior, S.D.; Odedra, S.; Erbil, M.A.; Karamanoglu, M.; Collins, R. Investigating the use of the Coanda Effect to Create Novel Unmanned Aerial Vehicles. In Proceedings of the International Conference on Manufacturing and Engineering Systems; ACM: New York, NY, USA, 2009; pp. 386–391.
15. Mirkov, N.; Rašuo, B. Numerical Simulation of Air Jet Attachment to Convex Walls and Applications. In Proceedings of the 27th ICAS Congress, Nice, France, 19–24 September 2010; Curran Associates, Inc.: Red Hook, NY, USA, 2010; pp. 1–7.
16. Mirkov, N.; Rašuo, B. Manuverability of an UAV with Coanda Effect Based Lift Production. In Proceedings of the 28th ICAS Congress, Brisbane, Australia, 23–28 September 2012; ICAS: Bonn, Germany, 2012; pp. 1–6.
17. Mirkov, N.; Rašuo, B. Numerical simulation of air jet attachment to convex walls and application to UAV. In *Boundary and Interior Layers, Computational and Asymptotic Methods*; Lecture Notes in Computational Science and Engineering; Knobloch, P., Ed.; Springer: Cham, Switzerland, 2015; Volume 108, pp. 197–208. [CrossRef]
18. Ameri, M. An Experimental and Theoretical Study of Coanda Ejectors. Ph.D. Thesis, Case Western Reserve University, Cleveland, OH, USA, 1993.
19. Kim, H.D.; Rajesh, G.; Setoguchi, T.; Matsuo, S. Optimization study of a Coanda ejector. *J. Therm. Sci.* **2006**, *15*, 331–336. [CrossRef]
20. Djojodihardjo, H.; Abdulhamid, M.F.; Basri, S.; Romli, S.I.; Abdul Majid, D.L.A. Numerical Simulation and Analysis of Coanda Effect Circulation Control for Wind-Turbine Application Considerations. *IIUM Eng. J.* **2011**, *12*, 19–42.
21. Gan, C.; Sahari, K.S.M.; Tan, C. Numerical investigation on Coanda flow over a logarithmic surface. *J. Mech. Sci. Technol.* **2015**, *29*, 2863–2869. [CrossRef]
22. Greitsch, L.; Eljardt, G.; Krueger, S. Operating conditions aligned ship design and evaluation. In Proceedings of the 1st International Symposium on Marine Propulsors, Trondheim, Norway, 22–24 June 2009.
23. Dumas, A.; Pascoa, J.; Trancossi, M.; Tacchini, A.; Ilieva, G.; Madonia, M. Acheon project: A novel vectoring jet concept. In Proceedings of the ASME 2012 International Mechanical Engineering Congress and Exposition, Houston, TX, USA, 9–15 November 2012; pp. 499–508.
24. Trancossi, M.; Dumas, A. *Coanda Synthetic Jet Deflection Apparatus and Control*; No. 2011-01-2590. SAE Technical Paper; SAE: Warrendale, PA, USA, 2011.
25. ANSYS Inc. *Ansys 14.0—FLUENT User's Guide*; ANSYS Inc.: Canonsburg, PA, USA, 2011.
26. Patankar, S.V. *Numerical Heat Transfer and Fluid Flow*; McGraw-Hill: New York, NY, USA, 1980.
27. Versteeg, H.K.; Malalasekera, W. *An Introduction to Computational Fluid Dynamics: The Finite Volume Method*; Pearson: London, UK, 2007.
28. Launder, B.E.; Spalding, D.B. *Lectures in Mathematical Models of Turbulence*; Academic Press: London, UK, 1972.
29. Wilcox, D.C. *Turbulence Modeling for CFD*, 3rd ed.; DCW Industries: La Cañada Flintridge, CA, USA, 2006.
30. Pinelli, A.; Naqavi, I.; Piomelli, U.; Favier, J. Immersed-boundary methods for general finite-difference and finite-volume Navier-Stokes solvers. *J. Comput. Phys.* **2010**, *229*, 9073–9091. [CrossRef]
31. Mirkov, N.; Rašuo, B.; Kenjereš, S. On the improved finite volume procedure for simulation of turbulent flows over real complex terrains. *J. Comput. Phys.* **2015**, *287*, 18–45. [CrossRef]
32. Digimizer. 5.3.4, MedCalc Software Bvba. 2018. Available online: www.digimizer.com (accessed on 30 July 2018).
33. Bradshaw, P. *Effects of Streamline Curvature on Turbulent Flow*; AGARDograph AG-169; AGARDograph: Paris, France, 1990.
34. Schlichting, H. *Boundary Layer Theory*; Mc-Graw Hill: New York, NY, USA, 1979.
35. Morgans, R.C.; Dally, B.B.; Nathan, G.J.; Lanspeary, P.V.; Fletcher, D.F. Application of the revised Wilcox (1998) k-turbulence model to a jet in co-flow. In Proceedings of the Second International Conference on CFD in the Mineral and Process Industries, Melbourne, Australia, 6–8 December 1999.
36. Georgiadis, N.J.; Yoder, D.A.; Engblom, W.B. Evaluation of modified two-equation turbulence models for jet flow predictions. *AIAA J.* **2006**, *44*, 3107–3114. [CrossRef]

37. Teixeira, F.B.; Lorenzini, G.; Errera, M.R.; Rocha, L.A.O.; Isoldi, L.A.; Dos Santos, E.D. Constructal Design of Triangular Arrangements of Square Bluff Bodies under Forced Convective Turbulent Flows. *Int. J. Heat Mass Transf.* **2018**, *126*, 521–535. [CrossRef]
38. ANSYS Inc. *Fluent, Ansys 12.0 Theory Guide*; ANSYS Inc.: Canonsburg, PA, USA, 2009.
39. Trancossi, M.; Stewart, J.; Maharshi, S.; Angeli, D. Mathematical model of a Constructal Coanda Effect nozzle. *J. Appl. Fluid Mech.* **2016**, *9*, 2813–2822.
40. Subhash, M.; Trancossi, M.; Pascoa, J. An Insight into the Coanda Flow Through Mathematical Modeling. In *Modeling and Simulation in Industrial Engineering*; Springer: Cham, Switzerland, 2018; pp. 101–114.
41. Mishra, A.A.; Iaccarino, G. Uncertainty Estimation for Reynolds-Averaged Navier Stokes Predictions of High-Speed Aircraft Nozzle Jets. *AIAA J.* **2017**, *55*, 3999–4004. [CrossRef]

 © 2019 by the authors. Licensee MDPI, Basel, Switzerland. This article is an open access article distributed under the terms and conditions of the Creative Commons Attribution (CC BY) license (http://creativecommons.org/licenses/by/4.0/).

Article

# Preliminary Study on the Contribution of External Forces to Ship Behavior

**Augusto Silva da Silva [1,*,†], Phelype Haron Oleinik [1,†], Eduardo de Paula Kirinus [1,†], Juliana Costi [2,†], Ricardo Cardoso Guimarães [1,†], Ana Pavlovic [3,†] and Wiliam Correa Marques [2,†]**

1. Escola de Engenharia, Universidade Federal do Rio Grande Campus Carreiros, Av. Itália s/n Km 8, Rio Grande 96201-900, Brazil; phe.h.o1@gmail.com (P.H.O.); ekirinus@gmail.com (E.d.P.K.); ricardo_guimaraes@hotmail.com.br (R.C.G.)
2. Instituto de Matemática, Estatística e Física, Universidade Federal do Rio Grande Campus Carreiros, Av. Itália s/n Km 8, Rio Grande 96201-900, Brazil; ju.costi@gmail.com (J.C.); wilianmarques@furg.br (W.C.M.)
3. Department of Industrial Engineering, University of Bologna, Viale Risorgimento 2, 40136 Bologna, Italy; ana.pavlovic@unibo.it
* Correspondence: ju.costi@gmail.com
† These authors contributed equally to this work.

Received: 15 January 2019; Accepted: 20 February 2019; Published: 20 March 2019

**Abstract:** Computational modeling has become a prominent tool to simulate physical processes for research and development projects. The coastal region of southern Brazil is very susceptible to oil spill accidents. Currently, oil is intensively transported in the region due to the presence of the Rio Grande Harbor, the Transpetro Waterway Terminal (Petrobras) and the Riograndense S/A Oil Refinery. Therefore, simulations under ideal navigation conditions for ships with potentially polluting loads are important because their use can reduce oil spills and toxic compound accidents in the environment. Therefore, the main objective of this work is to present a preliminary study of the contribution of external forces to a ship's behavior over a simulation period of 5 h. The methodology is based on the development of a numerical model using LaGrangian formalism and the calculus of variations, besides Maneuvering Modeling Group (MMG Model). The external forces considered were the wind acting directly on the ship, waves driven by wind, the rudder, the force acting on the hull, inertial forces, and seawater density. The results indicate that at the beginning of the simulation, the inertial forces were of primary importance for controlling the trajectory of the ship. After 5 h of simulations, the ship had completely changed its trajectory due to forces suffered by the ship, classified according to MMG Model.

**Keywords:** numerical model; SHIPMOVE; MMG Model; external forces

## 1. Introduction

Shipbuilding industry challenges have intensified the development of new operational configurations of high-speed ships for ocean-articulated convoys [1], as well as the phase approach of side discharge arrangement for tankers and fuel-supplying techniques to military vessels. These new operation conditions have produced additional challenges into the hydrodynamics interference on vessel dynamics mostly caused by radiated and diffracted waves between vessels. With that in mind, new resources aiming to better understand the vessel interaction with the ocean are constantly being developed, and with the advances of numerical models, these are, more often than not, being used to assess the operational limits and the effects of the hydrodynamic interference, of ship structures.

The influence of waves and hydrodynamic forces on vessel dynamics have been the subject of many studies over the years. Froude [2] conducted the pioneer study of ship movement, considering the effects of successive waves on ship movement. The study of Michell [3] was the first to take into account the disturbance imposed by a vessel in the flow, and originated the Thin Ship Theory. This theory was further developed and improved by many other authors (e.g., [4–6]). More recently, reference [7] presented the Band's Method, which evaluates the influence of hydrodynamic forces along bands of the vessel.

Then, a research group called Maneuvering Modeling Group (MMG) developed a numerical model to study the movements performed by a vessel when maneuvering. This model is known as MMG Model, and was reported in the Bulletin of Society of Naval Architects of Japan in 1977 [8]. However, the data used in this simulation were not detailed. Subsequently, Ogawa and Kasai [9] and Inoue et al. [10] published studies detailing this study, including equations describing the external forces.

Many simulations based on MMG Model were reproduced and published using different methods to acquire data, and manipulate and interpret results, so the Japan Society of Naval Architects and Ocean Engineers created a research committee to develop studies using a standard method for MMG Model. Yasukawa and Yoshimura [8] made a introduction of this standard method, splitting it in four essential elements and used a tanker to validate the study, concluding that the method used could represent the movements realized by a vessel and would be useful to predict these movements at full scale.

This study is about numerical simulations that use a specific model called SHIPMOVE. This numerical model uses variational theory and LaGrangian mechanics to calculate the trajectory of the ship and its angular movements.

SHIPMOVE does not use the theories already cited, but it incorporates in its basic equations the MMG Model equations. The SHIPMOVE basic equations describe the ship's position considering the kinetic and potential energy. Also, this model considers additional mass and drag forces due to friction and generation of waves.

Over the years, several methods have been developed to investigate the behavior of ships under the influence of external forces in seakeeping or maneuvering. Thus, the objective of this study is to present the first results obtained using a numerical model that was developed to study ship dynamics under the influence of external environmental forcing.

## 2. Material and Methods

To perform this study, numerical simulations were performed using the hydrodynamic model TELEMAC-3D, the wind-driven waves model TOMAWAC, both part of the open TELEMAC-MASCARET suite[1] and SHIPMOVE. The latter was developed by the *Laboratório de Análise Numérica e Sistemas Dinâmicos*[2]. The simulations were realized through the coupling of TELEMAC-3D+TOMAWAC feeding SHIPMOVE with hydrodynamic and wave data.

The SHIPMOVE model describes ship movement in the time domain using variational theory and LaGrangian mechanics. A detailed description of the first version of the model that does not include the MMG formulation, is provided in [11].

### 2.1. TELEMAC-3D

The open TELEMAC-MASCARET system is formed by a set of modules, operating in two or three dimensions, which can be used for different case studies such as hydrodynamics, sediment transport, dredging processes, waves driven by wind, pollutants, or oil spills, renewable energy, and more.

TELEMAC-3D is part of the open TELEMAC-MASCARET system, and it is responsible for the hydrodynamics of this study. It solves the Navier-Stokes equations [12] considering hydrostatic pressure and Boussinesq approximations [13]. Also, it solves the free surface variations in function of

---

[1] http://www.opentelemac.org
[2] www.lansd.furg.br

time. To solve the advection and diffusion hydrodynamic equations, the model uses the finite element techniques meanwhile for vertical discretization, TELEMAC uses sigma levels to segment the ocean from the surface to the bottom boundaries [13].

Some of the approximations used in TELEMAC include the simplification of the vertical velocity $w$, since it presents a small scale when compared with the $u$ and $v$ velocities. In this way, the vertical diffusive term in addition to other vertical terms are simplified from the calculations. However, terms that consider hydrostatic pressure variations and gravity acceleration are maintained in the vertical equation of momentum, so the pressure at any point in the ocean depends on atmospheric pressure above the water column weight over this point. Due to Boussinesq approximation ($\Delta \rho$ supposedly small when compared to a reference), it is possible to define the influence of the seawater state equation [14], because this equation relates specific mass of the fluid with the tracer concentration in this water.

Substances or properties that are present in water, such as temperature, salinity, and suspended sediments, are defined as scalar quantities, and can be active (when they interact with the hydrodynamic), or passive (when they do not interact). The advection processes, governed by sea current and wind, and diffusion, governed by turbulent processes from sources and sinks, control the temporal evolution of these scalar quantities.

In this way, TELEMAC-3D is used in the system to provide information about the hydrodynamics, mostly the current velocities, which will later feed the SHIPMOVE module.

### 2.2. TOMAWAC

To add the wave field into the vessel simulations, TOMAWAC (TELEMAC-Based Operational Model Addressing Wave Action Computation) module was used to perform the calculations. TOMAWAC is a third-generation wave model and part of the open TELEMAC-MASCARET suite.

TOMAWAC is a spectral model and generates waves through the wind, based on the wave action density conservation equation [15]. TOMAWAC considers most physical processes involved in wave generation, such as shoaling, whitecapping, dissipation through bottom friction, non-linear interactions, and depth-induced refraction; however, it does not consider wave diffraction and reflection [15].

To solve the equation cited above, TOMAWAC splits the direction wave spectra in a finite number of propagation frequencies and directions and solve these equations for all components without parametrization to any direction, spectral energy or wave action. Every wave spectrum component evolves in time according the modeled system.

The function of the TOMAWAC module in this study is to feed the SHIPMOVE model with ocean wave properties, such as the significant wave height and the peak direction of the waves.

2.2.1. Maneuvering Modeling Group (MMG Model)

MMG Model is one of the various numerical methods used to solve simulations involving ship maneuverability, and was used in this study. This model applies to ships, characteristics, and consequences from the involved forces at hull, propeller, and rudder, beyond the interaction between them.

The main objective of this model is to simplify the external forces application on vessel routes studies; however, there are some methods used on simulations that harm those studies, since one method used in some cases may be not appropriated to another one.

In this paper, 6 DOF are considered, differentiating from [16], where only surge, sway, and yaw movements were considered.

According to Maimun et al. [17], the ships have their movements significantly affected when sailing in shallow water, such as canals, because the hydrodynamic interaction between the ship and the bank located on laterals or bottom restrict the movements and develop other forces. Here, the bank effect will not be considered since the vessel travels to places where the relation between draft/depth does not produce this effect.

In this way, the force equations produced by the MMG Model, for both, $x$ and $y$ axis follows (Equations (1) and (2)):

$$X_A + X_W + X_R + X_H = F_x^e \tag{1}$$

$$Y_A + Y_W + Y_R + Y_H = F_y^e \tag{2}$$

where the subscriptions $A, W, R, H$ represent wind, wave, rudder, and hull, respectively.

2.2.2. Wind forces

For the wind forces that act on the vessel Equation (3) are presented. These equations are used by the SHIPMOVE module to simulate the action of the wind at the ship:

$$X_A = \frac{1}{2}\rho_A V_A^2 A_T C_{XA}(\theta_A) \tag{3a}$$

$$Y_A = \frac{1}{2}\rho_A V_A^2 A_L C_{YA}(\theta_A) \tag{3b}$$

where $\rho_A$ is the density of air, $V_A$ is the wind velocity, $A_T$ and $A_L$ are lateral projected area and frontal projected area, respectively. $C_{XA}$ and $C_{YA}$ are coefficients based on the relative wind direction, estimated by Andersen [18].

2.2.3. Wave Forces

For the wave effects acting on the vessel, Equation (4) are presented. The wave properties are first calculated trough TOMAWAC and with its output, the SHIPMOVE model is then capable of simulating the wave forces that are acting on the ship:

$$X_W = \rho g h^2 B^2 / L \overline{C_{XW}} \tag{4a}$$

$$Y_W = \rho g h^2 B^2 / L \overline{C_{YW}} \tag{4b}$$

where $\rho$ is the density of seawater, $g$ is the acceleration of gravity, $B$ represent ship's breadth, $h$ is the significant wave height, $\overline{C_{XW}}$ and $\overline{C_{YW}}$ are coefficients calculated by the Research Initiative on Oceangoing Ships (RIOS) [19].

2.2.4. Rudder Forces

SHIPMOVE also takes into account the forces provoked by the rudder of the vessel (Equation (5)):

$$X_R = -(1-t_R)F_N \sin\delta \tag{5a}$$

$$Y_R = -(1+a_H)F_N \cos\delta \tag{5b}$$

where $F_N$ is the rudder normal force, $\delta$ is the rudder angle, $t_R$ and $a_H$ represent the steering thrust deduction factor and rudder force increase factor, respectively.

The rudder normal force can be expressed as Equation (6):

$$F_N = \frac{1}{2}\rho \frac{6.13\lambda}{\lambda + 2.25} A_d U^2 \sin\alpha_R \tag{6}$$

where $\rho$ is the density of water, $\lambda$ is the rudder aspect ratio, $U_R$ was approximated to the ship's velocity, $A_d$ denotes rudder area and $\alpha_R$ is the relative inflow angle to rudder [20].

### 2.2.5. Forces Acting on the Hull

The hydrodynamic forces on the ship's hull and used in the SHIPMOVE model are given as Equation (7):

$$X_H = \frac{1}{2}\rho L^2 U^2 X'_H(v', r') - R(u) \qquad (7a)$$

$$Y_H = \frac{1}{2}\rho L^2 U^2 Y'_H(v', r') \qquad (7b)$$

where $r'$ denotes non-dimensionalized yaw rate by $r(L/U)$, and $v'$ represent non-dimensionalized lateral velocity $v/U$. $X'_H$ and $Y'_H$ are polynomial functions of $v'$ and $r'$. $R(u)$ is the ship resistance, which depends on residuary resistance coefficient, frictional resistance coefficient and the form factor [20].

### 2.3. Numerical Experiments

To study the influence of external forces acting on the ship, five simulations were performed. Each simulation represents a different external force described by the MMG Model: hull, rudder, wave and wind, and the fifth simulation was a case where the ship navigated without any external forces. These cases were performed seeking for differences between the trajectories of the ship in each case and how the forces affect the performance of the ship.

All cases were performed with the same initial conditions and the same simulation day: 5 April 2012. The climate characteristics of this day were chosen because, from meteorological data sets[3], it was noticed that the sea conditions, currents, and waves were at a low strength, allowing the ship to perform its movements without any extreme forces, avoiding peaks. The vessel navigated five hours in the open sea direction, departing from the south Brazilian coast.

### 2.4. Initial and Boundary Conditions

The definitions of initial and boundary conditions are essential to the initialization of the numerical model. They specify the physical characteristics of the vessel used in the simulation and the environmental conditions surrounding the ship (water temperature, salinity, water density, significant wave height, velocity, and acceleration of the waves).

The physic characteristics of the ship are shown in Table 1.

**Table 1.** Physical characteristics of the ship simulated in this study.

| Mass (kg) | Length (m) | Width (m) | Draught (m) |
|---|---|---|---|
| $2.87 \times 10^7$ | 173 | 32 | 12 |

The hydrodynamic simulation required the definition of boundary and initial conditions for the TELEMAC3D, which were obtained from the Hybrid Coordinate Ocean Model (HYCOM[4]). HYCOM data provides water temperature and salinity and current intensity with horizontal resolution of $0.083°$; such data is interpolated to the ocean boundary conditions.

At the free surface, the atmospheric boundary conditions (wind, air temperature, and pressure) were obtained from NCEP/NCAR Reanalisys[5], from NOAA, with spatial resolution of $1.875°$ and temporal resolution of 6 h.

---

[3] http://www.ecmwf.int/en/research/climate-reanalysis/era-interim
[4] https://hycom.org
[5] https://www.esrl.noaa.gov/psd/data/gridded/data.ncep.reanalysis.html

To initialize the wave model, TOMAWAC, WAVEWATCH III[6] dataset was used to force the ocean boundaries (spatial resolution of 30 min and temporal of 3 h). Wind data from NCEP/NCAR Reanalisys[7] with spatial resolution of 1.875° and temporal resolution of 6 hours were also used to force the superficial conditions of the wave model.

This study was carried out using data from the period between January and December of 2012. In order to represent the spatial domain, a mesh with 205.617 nodes (Figure 1) was used, with 1.5 km between nodes near the coast, and 8 km, approximately, in ocean boundaries.

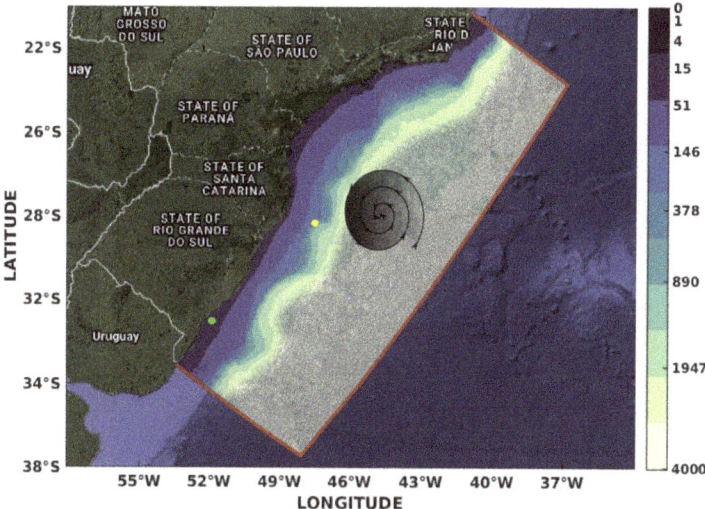

**Figure 1.** Study area located in the South-Southeastern Brazilian Shelf. The red lines represent the oceanic liquid boundaries and the spiral represents the atmospheric surface conditions. The green dot represents the initial position of the ship. The yellow dot represents the in situ data from buoy PNBOIA-SC. Bathymetry is shown in the color bar.

## 3. Calibration and Validation

To validate the hydrodynamic and wave models a simulation comprising the year of 2013 was performed and compared against observed buoy data.

A validation comparison was performed throughout a comparison of time series, the modeled and observed, and applied a set of performance skills and statistical metrics to validate TOMAWAC ($H_{mo}$ and $T_p$) and TELEMAC-3D (intensity of current velocity) results. The comparison was done with the time series measured by the wave buoys from PNBOIA[8] (Programa Nacional de Boias) which is a Brazilian project to collect oceanographic and meteorological data along the Brazilian coast. The data from PNBOIA were obtained on the coast of Santa Catarina (28.8333° S, 47.6000° W, 500 m depth) (Figure 1).

Despite the small number of observed data (5 months) available for 2013, the time series comparison between model and buoy data (Figure 2) indicates a good representation of the model results, when compared to the buoy data tendency. Nonetheless, visual comparison cannot be used on its own to define the accuracy of the model.

---

[6] ftp://polar.ncep.noaa.gov/history/waves
[7] https://www.esrl.noaa.gov/psd/data/gridded/data.ncep.reanalysis.html
[8] http://www.goosbrasil.org/pnboia/

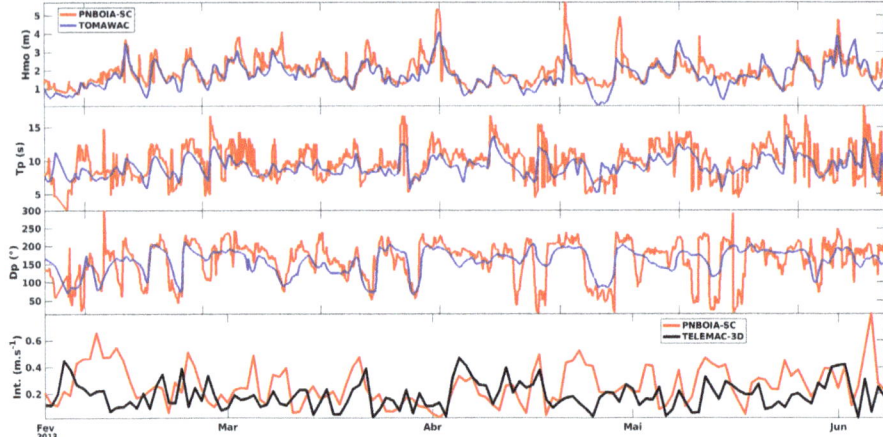

**Figure 2.** Time series of spectral significant wave heights (m), peak period (s) peak direction (°) from TOMAWAC (blue), and velocity intensity (m·s$^{-1}$) from TELEMAC-3D (black), compared against the observed data from the Santa Catarina PNBOIA buoy (red).

The usage of statistical metrics and well-known performance skills can quantify this comparison for better understanding the tendencies observed [21–25]. The desirable results for these statistical metrics are values closer to 0 (for RMSE, MAE, MSE, Bias, and SI), while others indicate better agreement between model and buoy with values closer to 1 (IC, SS, R$^2$).

At Table 2, $H_{mo}$ shows low RMSE values (near to 0.6 m), this tendency is enforced by low MAE values with 0.43. These values are lower than those found in the literature [23,26]. The Willmot Concordance Index (IC) [21] represents the ability of the model to reproduce the observations, resulting in 0.70, the accordance of the model data presents good confidence.

**Table 2.** Comparison between statistical metrics for wave and current data. Spectral significant wave height ($H_{mo}$), peak period ($T_p$) and current intensity (Int) analyzed among the Santa Catarina PNBOIA buoy. RMSE, MAE, Bias, $mod_{mean}$, $obs_{mean}$, $mod_{std}$, $obs_{std}$ are in the same units of the parameters, SI is in percentage. NA = Not Applied.

|             | $H_{mo}$ | $T_p$ | Int  |
|-------------|----------|-------|------|
| $mod_{mean}$ | 1.97    | 9.85  | 0.26 |
| $obs_{mean}$ | 1.75    | 9.10  | 0.18 |
| $mod_{std}$  | 0.70    | 2.53  | 0.14 |
| $obs_{std}$  | 0.67    | 1.40  | 0.10 |
| RMSE        | 0.57     | NA    | 0.16 |
| MAE         | 0.43     | NA    | 0.13 |
| IC          | 0.70     | NA    | 0.86 |
| MSE         | 0.33     | NA    | 0.02 |
| Bias        | 0.22     | 0.74  | 0.07 |
| SI (%)      | 33       | 23    | NA   |
| SS          | 1.12     | 1.10  | NA   |
| R$^2$       | 0.70     | 0.65  | NA   |

Despite the differences presented (Table 2), other wave statistics (R$^2$, Bias, SI and SS) were also applied to the analysis [22,27,28] to better validate the accuracy of the model. Khandekar [27] performed several model verifications with statistical methods, concluding that most models with SI lower than 35% are considered able to predict the studied environment; for instance, all SI comparisons (Table 2) show competence [28].

Standard deviation and mean values are within the range of acceptance, with bias close to zero indicating excellent agreement between predicted and observed data. A slight overestimation of the model data over the buoy is observed since SS values are higher than 1. The Pierson correlation coefficient (R) remains around 0.70, which is an acceptable result, according to Lalbeharry [23].

TELEMAC-3D ability to predict the buoy data was evaluated for the intensity of velocity (Table 2). RMSE (MAE) of 0.16 m·s$^{-1}$ (0.13 m·s$^{-1}$) presented low values in this comparison resulting in good agreement, in addition to an IC of 0.86. The MSE presented values closer to zero in all cases, indicating good agreement between model and buoy.

In this section a validation has been discussed in terms of current and wave parameters, whereas the hydrodynamic model TELEMAC-3D and wave model TOMAWAC were able to show good reproducibility of results behalf the buoy data and also being able to maintain the patterns and tendencies observed.

## 4. Results and Discussions

To study the influence of each force involved in the MMG Model methodology, five simulations were performed with the forces applied separately. The vessel had an initial velocity of 9.72 knots on the $x$ axis, and $-9.72$ knots on the $y$ axis, resulting in 13.7 knots approximately (the resulting vector). This energy was dissipated over the time according to the intensity of each force of the MMG Model and the dissipative forces.

The forces applied in the model are not necessarily dissipative forces only—they can be viscous forces, sometimes dissipating the ship energy, sometimes "helping" the ship to perform its movements. It depends on the relative direction between the forces and the ship.

Figure 3 shows the results of 5 h of simulation on 5 April 2012. This figure shows the routes performed by the ship with each influence applied separately, in all cases with the same initial conditions. Wave height is shown in the color bar. White vectors represent the wave direction while the black vectors represent wind direction.

To compare the simulated cases and see the differences between the ship with and without external forces, a simulation where no forces were involved and the ship had only the initial velocity condition was performed; in this way, the vessel dissipation of the energy only occurred trough the viscous and frictional forces acting as drag forces, as well as the dissipation trough the additional mass of the ship.

Chen et al. [19] showed the wind influence on a vessel's trajectory, when the vessel trajectory was deflected to North due to the influence of South acting wind. In the present study, it was possible to confirm that wind factor exerts a major role determining the trajectory distance and direction, while ocean currents induce more variations on rotational angles of the ship movements.

When the wind influence was simulated by the SHIPMOVE, it was possible to notice affinity to the results presented by [19]. On the simulation day, the wind conditions had major incidence direction of east. In this way, it was expected that the ship should deflect to starboard, and start to change its route direction South. From Figure 3 it is possible to notice that the ship did deflect to starboard corroborating with the previously studies, and demonstrating the accuracy of the SHIPMOVE.

Also, in Chen et al. [19], the effects of incident waves striking the lateral of vessels were studied. The authors conclude that waves focusing on the lateral of the ships have the strength to drive the vessels to the wave direction. For example, if the waves come from the South and collide on the lateral of a vessel, it will deflect the vessel trajectory a few degrees North.

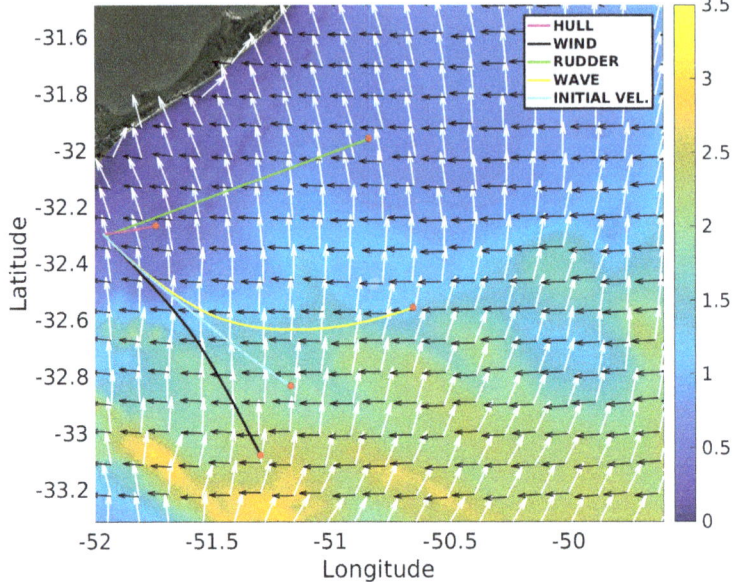

**Figure 3.** Routes performed by the ship in 5 hours of simulation.

Once again the present study demonstrated good agreement with the literature. For the case study, the waves of the simulated day were traveling North, with significant wave height varying from 2.5 m to 1.5 m close to where the vessel trajectory was calculated. With these conditions, the wave action was capable of deflecting the vessel trajectory port and deflect the trajectory that firstly started South-east to East and tendency to go North-east.

When the rudder force is analyzed trough the SHIPMOVE model it is possible to notice the proportionality of the force compared to the ship velocity; therefore, the force tends to reduce over time along with velocity. In the beginning of the simulation, the force is capable of making the ship change its direction to the North-east; however it is not strong enough to make the ship dissipate its initial energy totally; then, the vessel obtained a large trajectory, been the forces in $x$ and $y$ axis positives, in this case, first quadrant of a Cartesian grid.

The case in which the ship suffers hull force is similar to the rudder force case, but in the hull case the force is great enough to make the ship dissipate its initial velocity, and that is the reason the vessel trajectory is so small. In the first moment, the force is so strong that it can almost stop the ship, but when its velocity becomes small, the forces follow the velocity decay, allowing it to move, even with a small velocity. This is why the ship does not stand still over the 5 h of simulation.

The total distance performed by the ship in all cases can be discovered by calculating the resulting distance; Figure 4 shows the distance in each axis of the translational movements of the vessel.

In the case of only wind force acting on the ship, it covered approximately 65 km on the $x$ axis and 90 km on the $y$ axis, totalizing 112 km. With only the hull force, 21 km on $x$ axis and 4 km on $y$ axis totalizing 22 km. The study case only with wave forces was when the ship had his greater trajectory on $x$ axis, 132 km, and 48 km on the $y$ axis, totalizing 140 km. The ship navigated 117 km in the rudder case (110 km on $x$ and 40 km on $y$). When the ship was simulated only with initial velocity, without any of the forces applied, it navigated 100 km, approximately (79 km on $x$ and 61 km on $y$). Figure 4 also shows the ship's linear movements on $z$ axis (heave movement), which has much smaller magnitude in all cases.

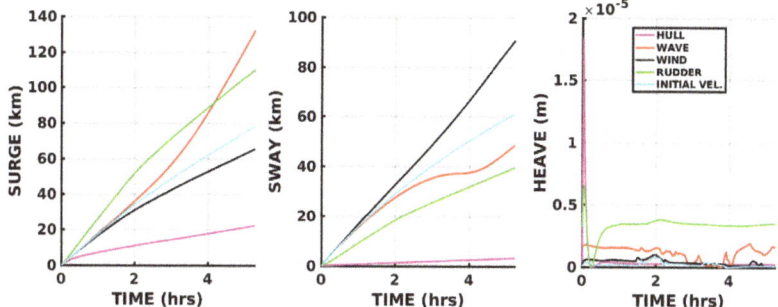

**Figure 4.** Translational movements performed by the ship.

The timeline of ship's velocity in each case can be seen in Figure 5.

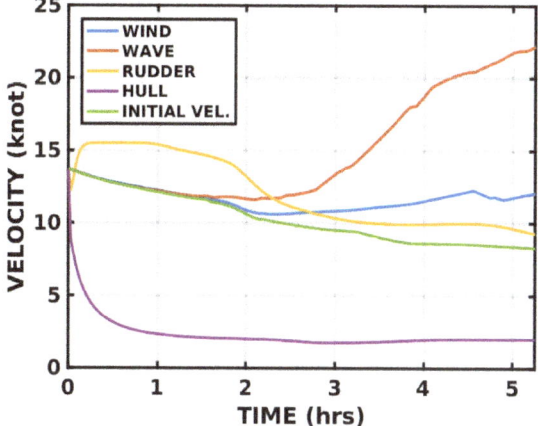

**Figure 5.** Timeline of ship's velocity.

To better understand the movements performed by the ship, we studied the velocity timeline of the ship over the 5 h of simulation for each case.

When the ship suffered only the wind forces, its velocity remained almost constant, reaching 10.5 knots near to the second hour of simulation, and then it raised to 12.5 knots approximately.

The ship had the greater velocity in the wave forces case, reaching 22.5 knots in the last hour of simulation, but it had a minimum (12.5 knots) between the second and third hour of simulation. The increase of the velocity can be attributed to the wave direction. In Figure 3, the wave starts to push the ship North.

In the rudder case, the ship had a peak near the beginning of the simulation, reaching more than 15 knots, but it starts to decrease after the first hour of simulation, and finishes with less than 10 knots. The hull case was the slowest case of them all. The ship began with 13.7 knots and then started to slow down from the beginning because the velocity depends on the force. The ship, after the first half hour of simulation, stabilized near to 2.5 knots and finished with this speed.

To finish, the case where the ship navigated only with the initial velocity dissipating its energy due to dissipative forces, had an initial velocity of 13.7 knots and, almost linearly, decreases to 8 knots. This is acceptable due to the fact that no external forces are involved, so the dissipative forces remain almost the same during all 5 h of simulation, since the ship and fluid properties did not change during the simulation.

Also, the angular movements of the vessel were studied in for every case, as shown in Figure 6.

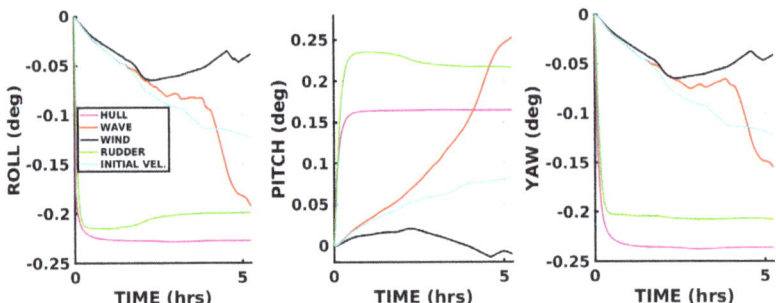

Figure 6. Rotational movements performed by the ship.

Looking at the roll movement, the vessel had the greater oscillation in the hull case (−0.225°), it can be attributed to the fact that the vessel had smaller speed, so the hydrodynamic forces could act more intensely.

On the other hand, in the wind case, the ship remained near to 0°, corroborating the study of Chen et al. [19] that said the wind has a smaller contribution on angular movements.

When the ship navigated without any external forces, its movement was a little larger when associated with the wind case, transcending −0.1°, but its variation from 0° to 0.125° was almost linear.

In the rudder case, the ship oscillated around −0.2°, with a peak (−0.22°) around the first hour of simulation. After that it stabilized on −0.2°.

The roll movement on the wave case was the case where the ship had the greatest instability, varying from 0° to almost −0.2°, and between the third and fourth hours of simulation the ship presented a movement of −0.075°.

Almost all cases were positive in the pitch movement, with the wave case as the largest one and the case where the ship had its greater variations, the same occurred in the roll movement, due to the high speed. It reached 0.25°.

Right below was the rudder case with 0.24° as a peak in the first hour. After this, it stabilized in 0.225° until the end. In the hull case the ship did not oscillate after the first hour, remaining on 0.16°. In the case without any external forces the ship had not transcended 0.1°, been almost linearly variation.

The only case in which this movement was negative was in the wind case, remaining near to 0° during the 5 h of simulation, showing again the fact that the wind has no major influence under angular movements.

In the yaw movement, the hull case was the case that the ship had the largest value, reaching −0.24°, but it had no oscillations during its simulation after the first hour, showing stability. The rudder case was the second largest value, remaining in −0.21° with a little variation between the first and the last hour of simulation.

When the ship navigated suffering no external forces, it had the second largest instability (0° to −0.125°), increasing almost linearly during its simulation. In the wave case, the ship had larger oscillations, varying from 0° to −0.07°, and then returns near to −0.05°. After the fourth hour of simulation, the ship performed yaw movements from −0.06° to −0.15° in the end. Again, the wind case was the case with fewer values related to angular movements. In general, all cases the ship had behaved similarly in roll and yaw movements, coincidentally.

When it is stated that there were more variations or instability, it means the ship had more variations during the 5 h of simulation—it does not mean the difference between the initial and the final value. It can be understood that in the cases where the hull and the rudder were studied, the ship reaches some stability near to the first hour and in the other cases, the angular movements vary during all simulation periods.

The behavior of the ship observed in all cases disagrees with the results reported by similar studies (Cha and Wan [29], Chuang and Steen [30], Bennett et al. [31], Ozdemir and Barlas [32]). Cha and Wan [29] had obtained regular sine-wave behaviors for the pitch movement, with linear characteristics. Such behavior was not observed on the Figure 6; however, the present study uses real waves and not regular waves such as in Cha and Wan [29].

The pitch movement angles presented by Ozdemir and Barlas [32] range between −3° and 3°, which is greater when compared to the angles found in the present study. However, the vessel used by Ozdemir and Barlas [32] was significantly larger (320 m), and the waves were 180° related to the ship's direction, while in the present study the waves focus with different angles.

It is crucial to acknowledge the uncertainties related to ship modeling in the ocean, due to the stochastic nature of this environment [33]. Such uncertainties are caused by the several approximations considered in both the hydrodynamic and ship model, the environmental data used to force the hydrodynamic model at its boundaries, and the MMG ship characteristics. The center of mass is a commonly used approximation in many ship models; however, it is expected that not considering the complex geometry of the ships may be an important source of uncertainty. Moreover, accessing the ship coefficients that are necessary to run and calibrate the MMG Model is still a difficult task, due to the private and commercial nature of the ships. Papanikolaou et al. [34] applied a model with 4 degress of freedom to simulate ship behavior under extreme conditions. They found that theorical and numerical results provide good results for practical uses, yet their accuracy decays as the simulation evolves over time.

## 5. Conclusions

Numerical simulations using three numerical models (SHIPMOVE, TELEMAC-3D and TOMAWAC) have been performed seeking to understand the movements performed by a ship under external forces classified trough the MMG Model. In this study, a few cases were simulated where each case corresponds to a different external force applied on the ship, allowing the authors understand the influence of each force.

In this way, some conclusions can be taken from the study:

- The ship had its trajectory deflect to starboard due to East winds, showing that wind has an influence on the ship behavior during navigation;
- The ship had its trajectory deflect to port due to North waves; in this way the waves can influence, significantly, the ship's route, mostly on the longer ones;
- The wind force has no major influences in the angles of angular movements, but it can influence the ship's trajectory mainly in longer distances;
- The wave force can significantly deflect the vessel trajectory and it is able to affectthe angles of angular movements performed by a ship;
- The external forces classified as MMG Model can push the vessel or make its linear movements difficult, which depends on the relative angle between the vessel and the forces;
- The external force direction associated with high intensity of sea and wave height can deflect a ship's trajectory or reduce its final displacement;
- The physical characteristics of a vessel, such as mass and geometric, form can influence the performance of a ship under severe scenarios;
- The velocity of a ship can significantly influence the external force magnitude, which depends on the relative velocity, including the hull and rudder forces;

From the results of this study, it can be concluded that external forces can affect significantly a ship's trajectory, and these effects can be observed in the final trajectory, and during the trajectory also as angular movements and the ship's instability.

As well as these results, the numerical modeling using the SHIPMOVE model can be an effective and useful tool for predicting and avoid unstable zones in the ocean, including vessel movement under various scenarios.

Moreover, the SHIPMOVE numerical model can be useful to study the influence of each external force and how researchers can improve the performance of a ship in severe scenarios.

**Author Contributions:** The individual contribution to this paper was performed throughout conceptualization, A.S.d.S., P.H.O., W.C.M. and E.d.P.K.; methodology, A.S.d.S., P.H.O. and J.C.; software, A.S.d.S., P.H.O. and W.C.M.; validation, A.S.d.S., P.H.O., W.C.M. and E.d.P.K.; formal analysis, P.H.O. and J.C.; investigation, P.H.O. and W.C.M.; resources, W.C.M.; data curation, A.S.d.S., P.H.O., W.C.M., J.C. and E.d.P.K.; writing—original draft preparation, A.S.d.S., P.H.O. and W.C.M.; writing—review and editing, R.C.G., J.C., A.P. and E.d.P.K.; visualization, A.S.d.S., P.H.O.; supervision, W.C.M. and E.d.P.K.; project administration, W.C.M.; funding acquisition, W.C.M.

**Acknowledgments:** The authors are grateful to the Conselho Nacional de Desenvolvimento Científico e Tecnológico (CNPq) under contract 146424/2017-4. Further acknowledgements go to the Brazilian Navy for providing detailed bathymetric data for the coastal area; the Brazilian National Water Agency, the PNBOIA project, NOAA, and HYCOM for supplying the validation and boundary conditions data sets, respectively; and the Open Telemac-Mascaret Consortium for freely distribution of the TELEMAC system making viable this research and finally a special thanks to the Supercomputing Center of the Federal University of Rio Grande do Sul (CESUP-UFRGS) and to the Sdumont Supercomputer from the Laboratório Nacional de Computação Científica (LNCC) (SDUMONT-2017-C01#166515) where most of the computational work was carried out. Although some data were taken from governmental databases, this paper is not necessarily representative of the views of the government.

**Conflicts of Interest:** The authors declare no conflict of interest.

## Abbreviations

The following abbreviations are used in this manuscript:

| | |
|---|---|
| CESUP-UFRGS | Supercomputing Center of the Federal University of Rio Grande do Sul |
| CNPq | Conselho Nacional de Desenvolvimento Científico e Tecnológico |
| HYCOM | Hybrid Coordinate Ocean Model |
| IC | Index of Concordance |
| JTTC | Japanese Towing Tank Conference |
| LANSD | Laboratório de Análise Numérica e Sistemas Dinâmicos |
| LNCC | Laboratório Nacional de Computação Científica |
| MAE | Mean Average Error |
| MMG Model | Maneuvering Modeling Group |
| $mod_{mean}$ | Mean value of the model data |
| $mod_{std}$ | Standard deviation value of the model data |
| MSE | Mean Square Error |
| NA | Not Applied |
| NOAA | (National Oceanic and Atmospheric Administration) |
| NCEP | National Centers for Environmental Prediction |
| NCAR | National Center for Atmospheric Research |
| $obs_{mean}$ | Mean value of the observed data |
| $obs_{std}$ | Standard deviation value of the observed data |
| PNBOIA | Programa Nacional de Boias |
| RIOS | Research Initiative on Oceangoing Ships |
| RMSE | Root Mean Square Error |
| $R^2$ | Pierson error coefficient |
| SI | Scatter Index |
| SS | Mean Square Inclination |
| SHIPMOVE | SHIP MOVEMENT MODEL |
| TOMAWAC | TELEMAC-Based Operational Model Addressing Wave Action Computation |

## References

1. Sclavounos, P. On the Diffraction of Free Surface Waves by a Slender Ship. Ph.D. Thesis, Massachusetts Institute of Technology, Cambridge, MA, USA, 1984.
2. Froude, W. *On the Rolling of Ships*; Royal Institution of Naval Architects: London, UK, 1862. p. 95.
3. Michell, J.H. Wave–Resistance of a Ship. *Philos. Mag.* **1898**, *45*, 106–123. [CrossRef]
4. Maruo, H. Calculation of the Wave Resistance of Ships, the Draught of Which is as Small as the Beam. *J. Zosen Kiokai* **1962**, *1962*, 21–37._21. [CrossRef]
5. Newman, J.N. A slender-body theory for ship oscillations in waves. *J. Fluid Mech.* **1964**, *18*, 602–618. [CrossRef]
6. Newman, J.N.; Sclavounos, P. The Unified Theory of Ship Motions. In Proceedings of the at Symposium on 13th Naval Hydrodynamics, Tokyo, Janpan, 6–10 October 1980.
7. Moreno, C.A. Interferência Hidrodinâmica no Comportamento em ondas Entre Navios com Velocidade de Avanço. Ph.D. Thesis, Coordenação de Aperfeiçoamento de Pessoal de Nível Superior, Brasília, Brazil, 2010.
8. Yasukawa, H.; Yoshimura, Y. Introduction of MMG standard method for ship maneuvering predictions. *J. Mar. Sci. Technol. (Jpn.)* **2015**, *20*, 37–52. [CrossRef]
9. Ogawa, A.; Kasai, H. On the mathematical model of manoeuvring motion of ships. *Int. Shipbuild. Prog.* **1978**. *25*, 306–319. [CrossRef]
10. Inoue, S.; Hirano, M.; Kijima, K.; Takashina, J. Practical calculation method of ship maneuvering motion. *Int. Shipbuild. Prog.* **1981**. *28*, 207–222. [CrossRef]
11. Armudi, A.; Marques, W.; Oleinik, P. Analysis of ship behavior under influence of waves and currents. *Revista de Engenharia Térmica* **2017**, *16*, 18–26. [CrossRef]
12. Hervouet, J.; Van Haren, L. Recent advances in numerical methods for fluid flow. In *Floodplain Processes*; Anderson, M.G., Walling, D.E., Bates, P.D., Eds.; Wiley: New York, NY, USA, 1996; pp. 183–214.
13. Hervouet, J.M. *Free Surface Flows: Modelling With the Finite Element Methods*; John Wiley & Sons: Hoboken, NJ, USA, 2007.
14. Millero, F.J.; Poisson, A. International one-atmosphere equation of state of seawater. *Deep Sea Res. Part A Oceanogr. Res. Pap.* **1981**, *28*, 625–629. [CrossRef]
15. TOMAWAC. *TOMAWAC Technical Report—Software for Sea State Modelling on Unstructured Grids over Oceans and Coastal Seas*; Technical Report; EDF: Chateau, France, 2011.
16. Lee, S.D.; Yu, C.H.; Hsiu, K.Y.; Hsieh, Y.F.; Tzeng, C.Y.; Kehr, Y.Z. Design and experiment of a small boat track-keeping autopilot. *Ocean Eng.* **2010**, *37*, 208–217. [CrossRef]
17. Maimun, A.; Priyanto, A.; Rahimuddin.; Sian, A.Y.; Awal, Z.I.; Celement, C.S.; Nurcholis.; Waqiyuddin, M. A mathematical model on manoeuvrability of a LNG tanker in vicinity of bank in restricted water. *Saf. Sci.* **2013**, *53*, 34–44. [CrossRef]
18. Andersen, I.M.V. Wind loads on post-panamax container ship. *Ocean Eng.* **2013**, *58*, 115–134. [CrossRef]
19. Chen, C.; Shiotani, S.; Sasa, K. Numerical ship navigation based on weather and ocean simulation. *Ocean Eng.* **2013**, *69*, 44–53. [CrossRef]
20. Zhang, W.; Zou, Z.J.; Deng, D.H. A study on prediction of ship maneuvering in regular waves. *Ocean Eng.* **2017**, *137*, 367–381. [CrossRef]
21. Willmot, C.J. Some Comments on the Evaluation of Model Performance. *Bull. Am. Meteorol. Soc.* **1982**, *63*, 1309–1313. [CrossRef]
22. Janssen, P.A.E.M.; Hansen, B.; Bidlot, J.R. Verification of the ECMWF Wave Forecasting System against Buoy and Altimeter Data. *Am. Meteorol. Soc.* **1997**, *12*, 763–784. [CrossRef]
23. Lalbeharry, R. Evaluation of the CMC regional wave forecasting system against buoy data. *Atmosphere-Ocean* **2002**, *40*, 1–20. [CrossRef]
24. Chawla, A.; Spindler, D.M.; Tolman, H.L. Validation of a thirty year wave hindcast using the Climate Forecast System Reanalysis winds. *Ocean Model.* **2013**, *70*, 189–206. [CrossRef]
25. Teegavarapu, R.S.V. *Floods in a Changing Climate: Extreme Precipitation*; Cambridge University Press: Cambridge, UK, 2013; p. 268.
26. Edwards, E.; Cradden, L.; Ingram, D.; Kalogeri, C. Verification within wave resource assessments. Part 1: Statistical analysis. *Int. J. Mar. Energy* **2014**, *8*, 50–69. [CrossRef]

27. Khandekar, M. Operational Wave Models. In *Guide to Wave Analysis and Forecasting*, 2nd ed.; World Meteorological Association: Geneva, Switzerland, 1998; Chapter 6, pp. 67–80.
28. Bidlot, J.R.; Holmes, D.J.; Whittmann, P.A.; Lalbeharry, R.; Chen, H.S. Intercomparison of the Performance of Operational Ocean Wave Forecasting Systems with Buoy Data. *Am. Meteorol. Soc.* **2002**, *17*. [CrossRef]
29. Cha, R.; Wan, D. Numerical investigation of motion response of two model ships in regular waves. *Procedia Eng.* **2015**, *116*, 20–31. [CrossRef]
30. Chuang, Z.; Steen, S. Speed loss of a vessel sailing in oblique waves. *Ocean Eng.* **2013**, *64*, 88–99. [CrossRef]
31. Bennett, S.; Hudson, D.; Temarel, P. The influend of forward speed on ship motions in abnormal waves: Experimental measurements and numerical predictions. *J. Fluids Struct.* **2013**, *39*, 154–172. [CrossRef]
32. Ozdemir, Y.; Barlas, B. Numerical study of ship motions and added resistance in regular incident waves of KVLCC2 model. *Int. J. Nav. Archit. Ocean Eng.* **2017**, *9*, 149–159. [CrossRef]
33. Papanikolaou, A.; Alfred Mohammed, E. Stochastic uncertainty modelling for ship design loads and operational guidance. *Ocean Eng.* **2014**, *86*, 47–57. [CrossRef]
34. Papanikolaou, A.; Fournarakis, N.; Chroni, D.; Liu, S.; Plessas, T. Simulation of the Maneuvering Behavior of Ships in Adverse Weather Conditions. In Proceedings of the 31st Symposium on Naval Hydrodynamics, Monterey, CA, USA, 16 September 2016.

© 2019 by the authors. Licensee MDPI, Basel, Switzerland. This article is an open access article distributed under the terms and conditions of the Creative Commons Attribution (CC BY) license (http://creativecommons.org/licenses/by/4.0/).

*Case Report*

# Environmental Management Systems and Balanced Scorecard: An Integrated Analysis of the Marine Transport

**Jelena Šaković Jovanović [1,*], Cristiano Fragassa [2], Zdravko Krivokapić [1] and Aleksandar Vujović [1]**

1. Faculty of Mechanical Engineering Podgorica, University of Montenegro, Cetinjska 2, 81000 Podgorica, Montenegro; zdravkok@ucg.ac.me (Z.K.); aleksv@ucg.ac.me (A.V.)
2. Department of Industrial Engineering, University of Bologna, viale Risorgimento 2, 40136 Bologna, Italy; cristiano.fragassa@unibo.it
* Correspondence: jelenajov@ucg.ac.me; Tel.: +382-(0)69-057-095

Received: 19 February 2019; Accepted: 23 April 2019; Published: 25 April 2019

**Abstract:** Critical aspects of the environment can reduce the efficiency of Environmental Management Systems (EMS) when applied to Marine Transport. Accordingly, this paper focuses on the improvement of the traditional EMS approach through the usage of Balanced Scorecard (BSC). The BSC represents a managing tool able to measure and increase organizational performance, taking into consideration environmental aspects. The proposed method, based on the ISO 14001 standard, allows management of environmental metrics through conventional BSC systems and it is applied to the biggest organization for marine transport in Montenegro as a case study methodology. In this qualitative investigation, particular attention was paid to creating EMS criteria able to orient the complete business operation of the organization but also to test their potential linkage to the conventional BSC approach. Four models of the BSC were created, each one including to a different extent the issue of environmental protection. Finally, an expert's evaluation of model efficiency, based on the ISO 9126, was carried out. As a result, the best ranked model is recommended for the selection of an approach toward environmental protection based on the use of the EMS metric in a conventional BSC system. This method—in short ECO-BSC—was developed for the specific benefit of those organizations operating on the marine transport market.

**Keywords:** marine transport; environmental management system; balanced scorecard; ISO 14001; ISO 9126; ISO 14598; AHP method; MCDM method

## 1. Introduction

The ISO 14000 group of standards provides recommendations and practical tools for companies and institutions of all types seeking to manage their environmental responsibilities. In particular, the ISO 14001, focused on the Environmental Management System (EMS), gives basic directions for the systematic improvement of environmental protection. Furthermore, it could be applied to all organizations regardless of their size, sector of activity and so on. EMS helps the firms to reduce their adverse environmental impacts while improving their economic efficiency [1]. The proof that an organization respects the requirements of ISO 14001 standards is a special certificate. The certificate compels an organization to pursue a way to improve their environmental management system; however, it is not a guarantee of real ecological enhancement. In other words, the ISO 14001 certification itself cannot ensure improvement of environmental performance.

In addition, it is noteworthy that the initial version of the ISO 14001, dated 2004, did not provide any direct obligation referring to the improvement of ecological goals. The new issue, dated 2015, includes conditions focused on how to improve ecological performance in management but achievements in that direction are still not mandatory prerequisites for obtaining the ISO 14001 certificate. As a consequence, even if this new issue is quite clear in showing the relevance of ecology in business, there is not a clear explanation for how an organization can measure and then improve environmental performance. This limit in the standard is maybe related to the fact that, up to now, practical results of specific managing situations, more than general methodologies, seem the most appropriate way to prove a positive effect of the implementation of an EMS.

Research articles focused on improvements of environmental performance through the implementation of the ISO 14001 standard, in fact, propose unresolved conclusions. On one side, Franchetti [2] is determined to show the positive effect of the application of environmental standards. In particular, the author examined the effect of the ISO 14001 certification on solid waste generation in US industrial organizations and found that solid waste generation rates are significantly reduced with EMS certification. On the other side, several groups of researchers propose discordant indications. In Reference [3–5] it was argued, for instance, that the ISO 14001 standard does not improve the environmental performance of organizations, while in References [6–8] quite the opposite evidence was provided, suggesting the possibility of performance declination.

Therefore, the ISO 14001 implementation should be amended in its application to include elements of performance management as the only way to ensure the continual harmonization with the requirements set out in the standard. According to this, a possible solution could be setting the ISO 14001 objectives throughout an entire organization and for all employees, through some instrument for performance measurement

In this sense, the Balanced Scorecard (BSC) as a strategic management system and performance management system could turn out to be quite a good choice for organizations that consider the environmental management system as strategically relevant.

In part 4 (four), different models of the BSC have been implemented and discussed, each integrating the same EMS orientations but in a different way. These models have been assessed by Experts with reference to a specific case study on marine transport. The evaluation has been performed in accordance with the requirements for product quality of the consolidated standard ISO 9126: Software engineering-Product quality (later replaced by the ISO 25010 with marginal changes [9]), widely adopted for the assessment of software products and similar applications.

The fundamental scope of the ISO 9126 is to correctly define priorities and assess them in reference to measurable values in accordance with four main directions: quality model, external metrics, internal metrics and quality in the use metrics. This quality approach takes into consideration general criteria:

- Functionality
- Reliability
- Efficiency
- Maintenance
- Portability
- Usability

These quality criteria are detailed by sub-characteristics (as *Suitability*; *Accuracy*; *Interoperability*; *Security*, ... ) and further divided into attributes. The most relevant peculiarities of those attributes are related to the fact that they have to be selected in a way that they represent a measurable and verifiable entity of the product. At the same time, those attributes are not rigidly defined a priori through the standard. Thus, the ISO 9126 can be reoriented for its use in the presence of different applications (in respect to the evaluation of software products), as in the present case.

The final assessment is performed in accordance with the procedure of evaluation defined in the standard ISO 14598 (later replaced by the ISO 25040 with minor changes in the procedure of evaluation).

## 2. Materials and Methods

### 2.1. Strategic Management System-Balanced Scorecard (BSC)

Among different concepts and tools for achieving high performance and, therewith, an efficient management system—such as SWOT (strengths, weaknesses, opportunities, threats) analysis [10,11], Value Chain [12], Lean Six Sigma [13], Total Quality Management [14,15] etc.—the Balanced Scorecard (BSC) represents a managing tool able to transform the strategy of an organization into objectives and measures (metric) at all organizational levels. Through BSC each individual in an organization has an opportunity to provide a personal contribution to strategy accomplishment and simultaneously has an insight into the value of the work.

Furthermore, the BSC identifies strengths in the company, highlighting in the analysed company the return over assets, return over equity, product quality, operative cycle time and the satisfaction of employees [16]. The strengths and weaknesses identified have a direct relationship with the achievement of goals, which are set according to the historical information of the company and the average values of the industry of stony aggregates.

The graphical presentation of the BSC in Figure 1, based on Reference [17], shows the translation path passing by the Mission, Vision and Strategy concepts into a system of activities grouped through 4 key Perspectives [18]. All perspectives, representing the direction of strategic activities inside an organization, are of equal importance and have to be monitored. In particular, these four perspectives must be continuously measured, analysed and improved so as to conduct comprehensive measurements [19]. This endless process of improvement was represented coupling a Top-Down direct path with a Bottom-Up feedback [20–22].

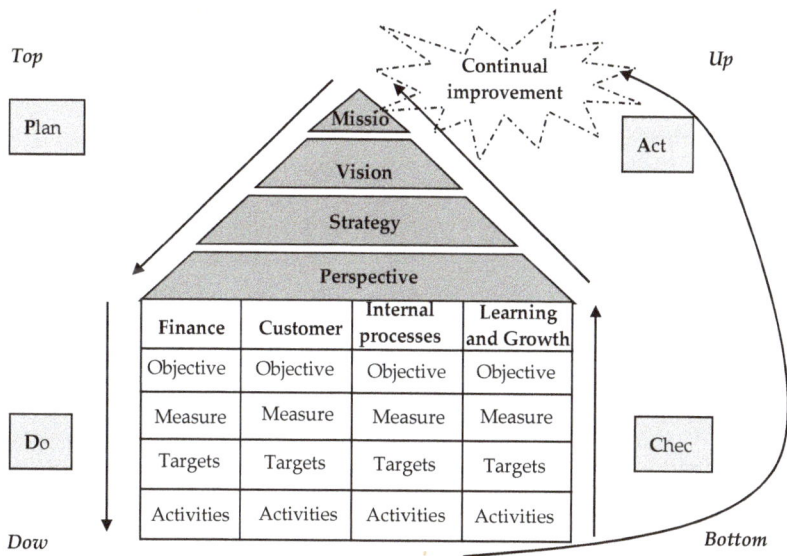

**Figure 1.** Managing mission, vision and strategy with the Balanced Scorecard (BSC) approach [17].

The importance of the BSC is particularly evident in the Bottom-Up path where the activities that are directed to the accomplishment of the target values are defined with the purpose of monitoring the degree of goal fulfilment. Such an approach continually provides management with a real picture of the organization's status in relation to defined strategic methods and indicates critical points where performance improvements are both possible and desirable. In order to attain the strategy, the objectives and measures of all defined perspectives have to be compatible and connected in a cause-and-effect chain so as to contribute to fulfilment of the strategy. In this way, the activity of an organization aimed at a continual improvement through the PDCA cycle becomes permanently initiated; the environment today is ever changing and therefore it is impossible for one defined strategy to be applied without improvements and reconsiderations.

The Balanced Scorecard recommends 4 standard perspectives (as listed in References [18,23,24]): learning and growth, internal processes, customers and finance; but these perspectives are not strictly defined parts that should be followed completely. Actually, the number and structure of perspectives depend on strategic organizational orientation, so some organizations can include in their strategy only 3 perspectives while other organizations can have 5 or more perspectives. Even though there is no strict form of the BSC system that could be applied to all organizations, it is possible to define two different concepts of the BSC which present the first decision in its implementation [7,20]:

- Concept of profit organization
- Concept of non-profit organization

Profit organizations are mainly focused on financial profit, which indicates that all objectives of the different perspectives are linked in a cause-and-effect chain, which is driven towards financial gain of the organizations. Non-profit organizations have a budget that services objective realization of all other perspectives, where the perspective of the customer's satisfaction is the most important and represents the outcome of the cause-and-effect chain of all other perspectives.

*2.2. Environmental Management System and Balanced Scorecard*

How can the metric of the environmental management system of a certain organization be included in the Balanced Scorecard system and which approach is more effective?

According to References [23,25–28] environmental and social aspects can be integrated into the BSC, in order to create a SBSC, in 3 different ways:

1. Modify the existing perspectives of the BSC model
2. Create new perspectives that include these elements
3. Create a special ecological/social scorecard

Between them, the most common approach is the first one (through 4 perspectives) but the second approach is also quite common, with the new perspective of focusing on environmental objectives. However, there are many supporters of the following concept: the addition of a new perspective depends on the strategic importance of these objectives [18,25,29–34].

However, EMS managers have a justified fear that there is an insufficient number of ecological objectives within the BSC model, meaning that these issues will be more and more neglected. It is also important to mention that this kind of involvement of the environmental objectives into a profit oriented BSC has to be realized in a way that also achieves the objectives of the financial perspective.

On the other side, in the BSC systems of non-profit organizations, the relations between perspectives are defined in order to achieve satisfaction of customers or stakeholders by using the available budget. According to this point of view, the non-profit approach of BSC could be much more efficient for the issue of an environmental management system than the profit oriented BSC model.

The third approach which includes the environmental metrics in the BSC represents the creation of the so-called SBSC (Sustainable Balanced Scorecard) which encompass all metrics that are in relation to the Environmental Management System, the social and financial perspective and which is connected with the BSC of the whole organization. The most important thing is that the SBSC is always profit oriented in profit oriented organizations. The possible negative aspect of its usage in practice is the establishment of a parallel system focused on EMS in relation to the conventional BSC. In accordance with the EMS established by ISO 14001, any other management system which is not entirely included in the BSC can also be considered a parallel system.

In line with this, there is a proposition where:

*The Balanced Scorecard system with the ECO BSC system based on the principal of non-profit organizations and oriented toward the environmental protection management system presents an efficient management system which mainly improves ecological performance.*

This hypothesis suggests the fact that the key to accomplishing an effective and efficient management system providing and improving environmental performance lies in a good connection between ECO BSC and the conventional BSC system where is ECO BSC is based on a non-profit model of connected perspectives.

### 2.3. The Case-Study: A Profit Organization Performing Marine Transport

In the paper "AD Barska plovidba", a profit organization located in Montenegro—an active marine transport organization—was studied as a case of the implementation of the EMS approach based on the conventional BSC system. This action allowed environmental metrics to be included in the company management for the first time with results used to evaluate and select the best model for implementation.

Probably better known as "Montenegro Lines", the Company was founded with the aim of providing marine transport services, marina activities, international freight forwarding and a marine agency. The Company carries out its basic activity (marine transport) through the transport of passengers and cargos (Figure 2) on the lines between Montenegro and Italy (mainly Bar-Bari-Bar). It is located in the Port of Bar which is traditionally seen as the strategic node in the pan-European corridor connecting East Europe and the Balkans (as Budapest, Belgrade) with Central-Southern Italy (Bari, Taranto) and other Mediterranean ports, relevant in terms of passenger and merchandise flows.

**Figure 2.** Montenegro Lines passenger ships and cargos at anchor in the Port of Bar, namely, "Sveti Stefan", "Sveti Stefan II" and "Bar"cargo (from left to right).

The Company conducts its core business with 2 passenger ships:

- "Sveti Stefan", class "Bureau Veritas", length 109 m, speed 21 knots, with the capacity of 550 passengers, 240 beds in the cabins and freight capacity of 205 vehicles,
- "Sveti Stefan II", class "Det Norske Veritas", length 118 m, 22 knots, with the capacity of 920 passengers, 495 beds in the cabins and freight capacity of 36 trucks (up to 12 m) or 24 (up to 18 m) or 225 cars.

Inside this organization, the conventional BSC system allowed planning for the entire business but prior to this survey, the issue of the environmental management was covered by a rather insignificant set of measures. At the same time, an established Safety Management System (SMS), based on the International Safety Management (ISM) Code and mainly oriented towards environmental protection, is prescribed by the International Maritime Organization (IMO) as a mandatory regulation for all marine companies operating in international maritime traffic.

This combination made "Montenegro Lines" a very interesting case study for the current analysis, also taking into account the new company strategy, which focuses on environmental protection, in due accordance with the international standards and the legislative requirements applicable to marine transport organizations.

## 3. Development

### 3.1. Approaches to Integrating the EMS into the Balanced Scorecard

Popular books about Balanced Scorecard [20,23,25,32] have not elaborated in detail the issues of implementation of the ecological aspects in BSC models; besides, they have presented strategic maps of specific for-profit organisations with a considerable number of ecological goals within existing BSC perspectives.

The environmental protection goals within these profitably oriented strategic maps (in the way that they lead to the fulfilment of financial goals as the organisation's ultimate goal) are mainly focused on the offer of a "green" product to sensible "green" buyers and on their financial gain.

Nevertheless, there are research papers on the sustainable concept of the BSC, namely SBSC as Sustainability BSC.

This concept primarily seeks to determine the strategically relevant environmental and social goals of the organization that lead to the creation of economic value by the causal and consequential relationships of the perspective so that it is oriented towards three key elements: finance, ecology and society.

Literary sources show that adding a new perspective or creating a special scorecard actually depends on the strategic relevance of environmental protection for the organisation [23–25,29–34].

### 3.2. Integration of Environmental and Social Aspects

Environmental and social aspects can be considered by the existing 4 perspectives through strategic elements, objectives and measures. In this way, environmental and social aspects become an integral part of the conventional BSC and are automatically integrated into a causal chain that is hierarchically oriented towards the goals of the financial perspective in profit organizations. These aspects need to be integrated into the market system and to tend to the 'green' products' buyers.

However, the number of environment-centred objectives and measures that involve such an approach in the BSC is fairly limited and insufficient to cover the overall issue.

However, a large number of organisations that are insufficiently focused on the issue of environmental protection still accept these approaches explaining that this issue is still covered by certain BSC objectives.

### 3.3. Creating a New Ecological/Social Perspective

With this approach, there is a large number of environmental and social aspects that cannot be fully integrated into the market changes for the simple reason that they are not being explicitly marketed. Creating a new perspective provides a clearer picture of the integration of environmental and social aspects into the conventional BSC model, at the same time creating an opportunity for expanding the metric that encompasses this issue. Elements of an ecological/social perspective must be linked to all other perspectives, not just to the financial perspective. It is also possible to create two perspectives that include independently social and ecological requirements.

However, in practice, the most common approach is the inclusion of sustainable development elements in the conventional BSC model, while adding the perspectives that these goals are unifying is less common because, for most organizations, sustainable development is not strategically relevant. This is the reason why there are numerous advocators of the concept that the addition of a "new perspective depends on the strategic relevance" of these elements [18,23,25,30–33,35,36].

### 3.4. Creating A Derived Ecological/Social Scorecard

The previous two concepts always create a justifiable concern with the managers responsible for environmental protection, on the grounds that this problem will be even more neglected due to a scarce number of objectives focused on the financial perspective within the BSC model.

The third approach to integrating environmental and social aspects into the BSC involves creation of a specific ecological/social scorecard in parallel with the conventional BSC. The derived ecological/social scorecard is not detached from the conventional BSC and must be associated with it in order to strengthen the ecological commitments of the organization's development.

Actually, scientists developed an extended architecture of BSC under the name of SBSC and they did it for two reasons [37]:

1. to allow management to define specific metrics in all the three dimensions of the company sustainability: economic, environmental and social;
2. to let SBSC merge these three dimensions in a single integrated management system instead of a parallel system.

The point is that this particular scorecard that would manage to cover the overall issue of sustainable development would be profit-oriented. The reason for the rare usage of this model in practice is the creation of a parallel system in relation to the conventional BSC [38,39].

However, in the literature, several examples of the application of SBSC exist but always dealing with EMS in profit-oriented situations [35,36,40,41].

Furthermore, in Reference [42], the authors claim not to have great expectations for SBSC, since this method seems not appropriate to permit radical changes in business operations when oriented at providing a sustainable development. In these terms, it is considered, in practice, a simple managing tool for the implementation of a strategy (as in References [42,43]). This is why the management has to set strategic priorities for sustainable development in the first place and then use SBSC for implementing strategic priorities [42]. The survey [26] is in compliance with the aforesaid as it says that the sustainable development must address a strategic direction of an organization, so as to provide the success of SBSC by continuous monitoring of metrics related to these issues and the continuous work of all employees.

Otherwise, sustainable development becomes an activity which is not considered a factor of competitive advantage for an organization and is therefore not dealt with in the organisation [26].

In Reference [44], an investigation implemented inside the port industry, it is shown how tools like BSC can permit employees to effectively take part in the EMS application; this condition represents an aspect fundamental for gaining environmental performance in real situations.

In Reference [45] it is pointed out that in all organisations where the clear compliance with the regulations governing environmental protection is mandatory, as in the case of port industry organisations, it is possible to combine different management tools with BSC with the aim of increasing the environmental protection effectiveness and efficiency. In this way, the application of management innovations would enable both employees and management to actively and efficiently find ways to work in the field of environmental protection [45].

In Reference [46], an assessment of the effect of different environment management strategies based on the Port of Alexandria, all environmental aspects of the port were presented in the first place for relevance. Then, by modelling the system, it appeared that the application of SBSC system, which provides *"more environmentally, socially and economically friendly approach to port operations over the long run"*, could effectively lead to the reduction of gas emission.

In Reference [46], in particular, it is also claimed that the SBSC approach is especially suitable for those organisations that have not implemented a BSC yet but want to integrate sustainable development goals into employees' everyday activities without establishing BSC for complete business practice. It can also be appropriate for those organisations with established BSC but wishing to get strategically oriented to and put special emphasis on the sustainable development [46].

In Reference [47], a survey is reported, done with regards to the relationship between the SBSC and a traditional eco-efficiency analysis. The study highlights that the *"eco-efficiency is an instrument for estimating and controlling the appropriate key performance indicators for two major aspects of sustainability; namely, environmental and economic issues"*. In this terms, the eco-efficiency analysis may be considered as a connection between SBSC and the EMS, just because both systems rely on material and energy flow analysis and use life-cycle assessment approaches.

Hence, experiences in the application of SBSC are quite different. However, it is typical that these approaches, as the ones previously presented, move toward the inclusion of sustainable development goals and measures into the BSC for profit organisations. Therefore, the environmental protection involves goals and measures connected through BSC perspectives in such a way to be completely in compliance with models for profit organisations. It means that they aim at fulfilment of financial perspective goals, as clearly shown in Reference [18,20,25,29,31,47–51].

In particular, [36] points out that specific conceptual modifications inside the BSC (respect to the original model) have enabled its evolution with the effect to increase its application in practise. The paper creates a new ECO BSC model respecting the operating principle of non-profit organisations but also considering the budget as a relevant issue inside all ecologically relevant measures. It means the attention to the budget is not limited only on those aspects that increase organisation's profit. At the same time, such model locates financial perspective on the top, in the way that financial effects of such financing of environmental protection can be measured. With this approach, tested in practice, new version of BSC, conceptually oriented to the environmental protection, can be created.

## 4. Results

*4.1. ECO BSC System Created on the Concept of a Non-Profit Organization*

With the aim of analysing all the approaches adopted of the EMS by the application of the BSC and with the aim of finding the most efficient way, the definition of 'ECO BSC' systems was initiated.

Objectives and measures of the ECO BSC are integrated within the framework of 5 perspectives (budget, learning and growth, internal processes, stakeholders, finance) and mutually connected in a cause-and-effect chain through strategic map created on the approach of BSC for non-profit organizations, as shown in Figure 3. Therefore, the essence of this model is to link the objectives and measures related to EMS so as to establish their mutual correlation and finally their impact on financial results. The point of such a model where objectives and measures are defined at all levels of the organization is that all employees at all levels of the organization are engaged in environmental issues and that such activities can be measured.

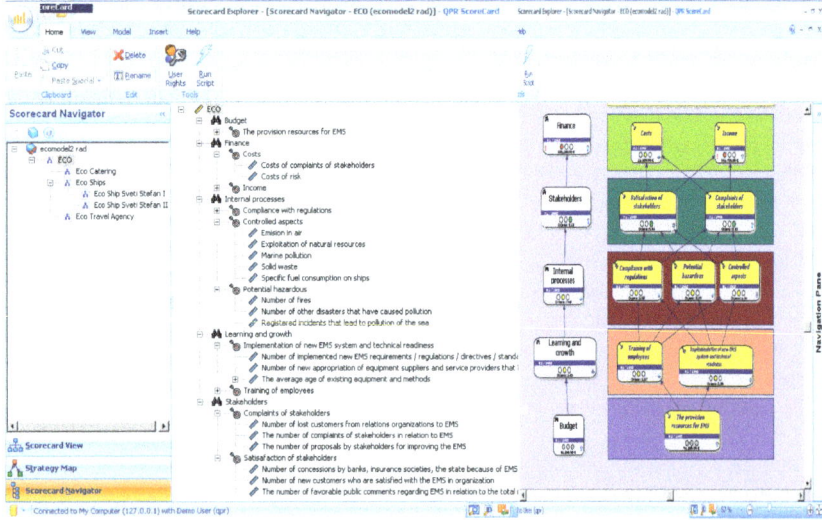

**Figure 3.** The ECO BSC system of the organization with a review of a strategic map (by QPR software).

Having in mind that the BSC system consists of a well-designed metrics that cascade translate organization's strategic objectives to all employees, for the purpose of creation of the ECO BSC system, the communication with the employees and the EMS manager was accomplished in this work. Thus, a list of goals and measures was identified, acting as an environmental metric at a corporate level, that depicts the strategic activities of this organization in the area of the EMS. After defining the entire metric, the objectives need to be translated to the lower levels in the organization in cascade-like manner, to reach all the employees and other resources.

Figure 3, shown in line with Reference [41], shows evidence of a portion of the features characterizing the ECO BSC system for the entire organization. This figure shows the general review of the organization in terms of a functional composition and strategic map, as expressed by the software suite QPR. Specifically, the layout splits the system information into three different frames portraying, from left to right: the overall structure with its substructures, list of (several) system features under consideration, their links and mutual influences.

In particular, left frame in Figure 3, it can be noted that the ECO BSC can be regarded as being comprised of 3 separated and ecologically oriented sub-systems, namely:

- ECO *Catering*
- ECO ships, specifically, Eco ship *Sveti Stefan I* and Eco ship *Sveti Stefan II*
- ECO *travel agency*

It is also evident—see the central frame in Figure 3—that the system, together with its subsystems, have been analysed taking into account traditional features such as: budget, finance, internal processes and such. But also additional aspects, less common in this kind of investigations, as potentially hazardous, stakeholders (complaints and satisfactions) have been included in the study.

This ECO BSC system is created on the approach of BSC for non-profit organizations as a first step; it means that the perspective budget and other similar perspectives are directed to the satisfaction of the stakeholders, instead of generating profit. It is not possible to act differently, since, in strictly financial terms, environmental protection would always be a cost. Any business management tool developed in exclusive profit logic would lead to the exclusion of the environmental protection measures, even at the cost of leading the company to act outside the limits set by the regulations. At the same time, the deflection of a business-oriented concept from the classic concept of non-profit organizations means creation of the fifth—financial perspective in order to measure financial effects of this kind of ECO BSC.

The solution is the one suggested below. The financial perspective in the management system under development does not represent the orientation of the BSC system but only depicts the financial consequences of the EMS in this way.

The ECO BSC can exist independently and cover the overall segment of the Environmental Management System and as such with a purpose of monitoring the entire EMS with measurements and enhancement of environmental performance.

To avoid creating a parallel management system with divergent goals, it is necessary to include a key metric (goals and measures) of the ECO BSC in the conventional BSC system.

This link can be realized in one of the following ways:

1. By adding one or more objectives from the ECO-BSC to the perspective of the conventional BSC;
2. By creating the fifth perspective with key objectives from the ECO-BSC;
3. By combining the two approaches above.

In this way, the ECO BSC would include entire environmental management system coordinated by an EMS manager while the BSC system would be coordinated by top management which is focused on the whole business with the key metric of the environmental management system.

*4.2. Approaches of Making Connection between ECO BSC and Conventional BSC*

As explained, the ECO BSC needs to be well connected with the conventional BSC model in order to avoid establishing parallel management systems

Aimed at establishing good connection between the ECO BSC and the conventional BSC model, the ECO BSC key metrics referring to the EMS need to be introduced into the conventional BSC. This is possible since each objective or measure in the scorecard framework (map of the cause-effect metric of one organizational level) can present functional dependency of other objectives or measures of the same or other scorecard.

Figure 4 shows connection between the ECO and the conventional BSC. Actually, all the perspectives of conventional BSC involved some EMS objectives and measures which established connections between the conventional BSC and the ECO BSC (for example: EMS process into internal processes of scorecard Ships, as represented in Figure 4). Following this, Model 1 was obtained and implemented. In particular, the left frame in Figure 4 shows how the system under investigation is changed in a way that considers more articulated situation. Instead of three subsystems (displayed in Figure 3, left frame), where only one of them—namely *EcoShips*—showed a further level of structure, here the overall model is initially divided in two branches: one attributable to the company as a whole and the other to the sustainability part of the business. Also in this case, features were presented, as shown in the central frame but they assumed different relevance and mutual connection in reference to the previous case.

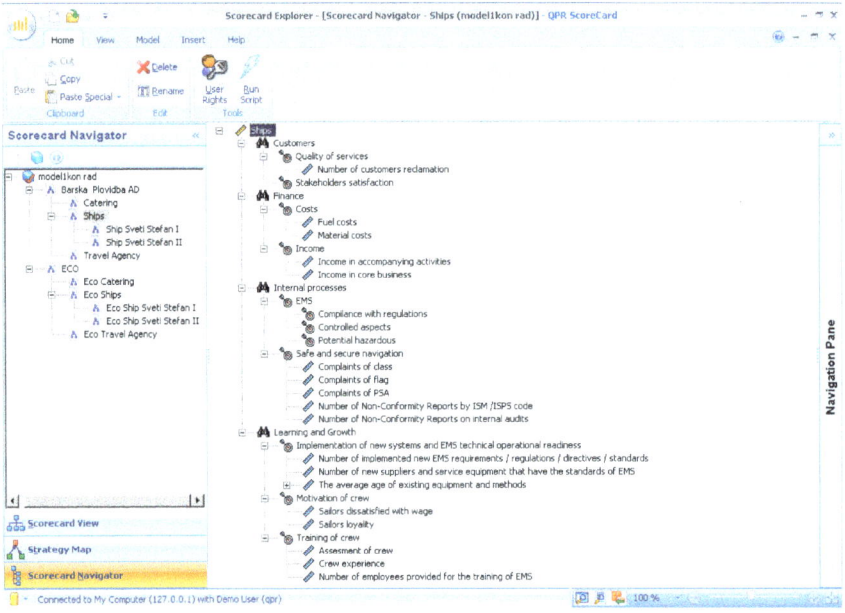

**Figure 4.** Model 1: Connection between conventional BSC and ECO BSC by involving key environmental objectives and measures into existed perspectives of conventional BSC model.

Another way the ECO and the conventional BSC system connection can be integrated, as listed, is when the fifth perspective, the so-called ECO perspective, is added to the conventional BSC system. This perspective includes key metrics of the ECO BSC, as shown in Figure ?? and allows creation of Model 2.

In accordance with literature sources [23,25–28] apart from the two previously designed models, additional two were created combining the previous approaches:

1. Model 3, which corresponds to the inclusion of the EMS metric in the existing BSC: this model is similar to conventional BSC of Model 1 but without the ECO BSC;

2. Model 4, which corresponds to the inclusion of the EMS metric in the newly created ECO perspective: this model is similar to conventional BSC of Model 2 but without the ECO BSC.

As a result, 4 different models have been developed that are able to include, in different ways, the aspects of environmental protection inside the conventional BSC system.

Since, for a matter of fact, it is complicated to analyse the efficiency of those alternative models in real working conditions, especially considering the quite long period (up to a decade) that would be necessary for the implementation of each model and for the evaluation of their quality, the choice of the most suitable one to be used in Montenegro Lines was realized based on expert evaluation. In this way, the risk of adopting an ineffective model has also been reduced.

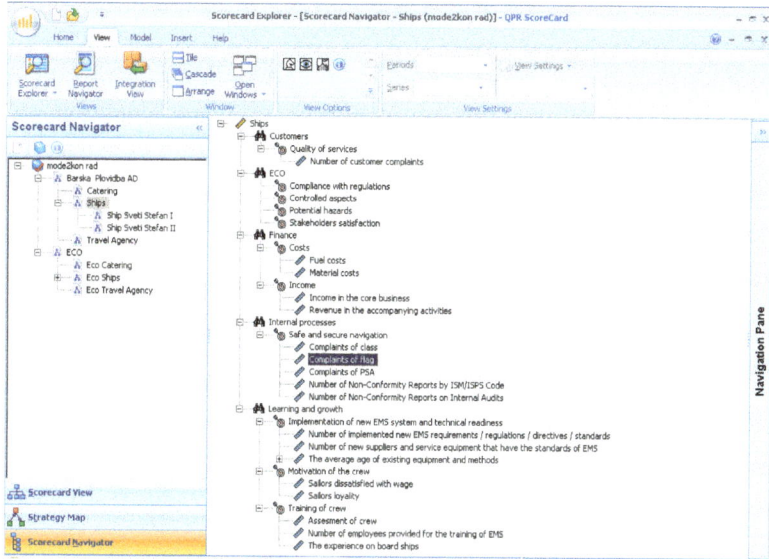

**Figure 5.** Model 2: Connection between the conventional BSC and ECO BSC by a new perspective called ECO.

## 5. Discussion

### 5.1. Model Assessment

With the aim of carefully implementing this process of model assessment, available standards were followed and, further to this, experts from the BSC and EMS area were asked to provide special support.

The standard *ISO/IEC 9126-Software engineering-product quality* is used for the assessment of the BSC models. The standard consists of criteria for assessment of internal and external quality of software products. According the standard, the assessment of the four BSC models was conducted on internal and external quality evaluating 6 criteria (Functionality, Reliability, Efficiency, Sustainability, Portability and Usefulness) and their 27 sub-criteria. Evaluation of the software (models) quality in use was postponed after a certain time of application of the best ranking model.

The model evaluation process was conducted by two teams of experts. Experts have been selected with the aim of providing the highest level of expertise in each of the specific areas covered by these models but also to ensure the maximum objectivity during the evaluation process.

Specifically, *Team 1*, consisting in academic members, was created with the scope to provide competences respect to the wide scientific and research knowledge on the topics covered by the models. In a complementary action, *Team 2*, consisting in employees and top managers, was created with the scope to provide a wide and deep practical knowledge about the key processes characterizing the organisation; furthermore, *Team 2* is also in charge of an evaluation on the application of new methods and technologies for the purpose of innovation and for the improvement of work processes in organisation: exactly the aspects that the present investigation has a chance to witness.

Table 1 summarizes the fields of expertise for the external experts and roles of the internal managers involved in the evaluation.

**Table 1.** Role and fields of expertise for External Experts (Academic Group) and Internal Experts (Company Group) involved in the assessment.

| External Experts (Team 1) | Internal Managers (Team 2) |
| --- | --- |
| Quality and ISO standards | Environmental Protection |
| Environmental Management Systems | Financial & Accountability |
| Performance measurement | Technical & Technological |

Entering in further details:

(1) *Team 1* experts are university professors and researchers, recognised for their references in those areas that are crucial for performing an evaluation on the considered issues, such as:

- Environmental Management System.
- Quality and ISO Standards.
- Performance measurement.

Their task during the analysis was focused on observing the importance of the models under investigation from the standpoint of scientific soundness and validity. The team conducted the evaluation in order to find the best solution for establishing an efficient and effective management system, respect to the whole organization, which can enable continuous measurement and improvement of environmental performance.

(2) *Team 2* experts are key-member inside the 'Montenegro Lines' organisation chart. As a result of their technical competences and long-time employment, these employees have relevant roles (e.g., responsible or managers) in fundamental processes. Their position and experience provide them a deep knowledge about company and work processes, also permitting an access to all available information. Hence, Team 2 is comprised of the following experts:

- Environmental Manager in charge to compliance with legal regulations and relevant environmental standards in the organization;
- Financial Director in charge to manage the organisation from the aspect of finances and accountability and fully aware of all financial aspects;
- Technical Director to charge to directly manage all processes performed inside the organisation, especially in terms of technical and technological aspects, with full access to the related information.

Their task during the analysis was focused on observing the importance of the models under investigation from the standpoint of practical applicability, effectiveness and efficiency respect to the processes performed in "Montenegro Lines." The team also assessed which model could fulfil the present and future company needs, especially in terms of strategic organizational management and environmental management.

Experts from both sides provided an evaluation based on their knowledge but also based on direct and indirect analysis on data from metrics, as well as by semi-structured interviews.

The evaluation was done in respect to the criteria reported to in Table 2, where each aspect used in classifying the internal and external evaluation items was reported. It represents the list questions the experts were asked to reply, able to represent the mentioned (6) criteria and (27) sub criteria for an accurate evaluation. The (6) experts performed their assessment independently from each other.

Table 2. Internal and external measures in relation to the 6 evaluating criteria.

| Criterion | Sub-Criteria | Internal Measure | External Measure |
|---|---|---|---|
| Functionality | Adjustability | Has the correct functionality been activated correctly? | What is the stability of the operations at the stage of the operation? |
| | Accuracy | How accurate are the data? | Do you get imprecise data at work? |
| | Security | How do I access the system and disable data manipulation? | How often is data manipulation occurring? |
| | Interoperability | Are there good connections between data? | How good is the exchange of data during work? |
| | Compliance | Have compliance with applicable regulations & standards achieved? | Have compliance with applicable regulations and standards achieved? |
| Reliability | Availability | How many mistakes have been detected and corrected product? | How many problems can there be that can cause mistakes in future work? |
| | Errors Tolerance | How much software does incorrectly define connections? | How often systems fail in work or are inability to answer the questions? |
| | Reparability | How effective is the system after some unforeseen operations? | How much system is able to answer in working on all the required questions? |
| | Compliance | What is the compatibility of the reliability of the system with the applicable regulations and standards? | What is the compatibility of the reliability of the system with the applicable regulations and standards? |
| Usability | Understandable | How easy is the connection for users to use? | How many users correctly use defined functions? |
| | Easy learning | How is the function clearly explained to users? | How many users need time to learn all the functions of the software? |
| | Operability | How many functions can be changed during operation? | Can the user easily enter the data? |
| | Attractiveness | How attractive is the interface? | How attractive is the interface? |
| | Compliance | Have compliance with applicable regulations and standards been achieved? | Have compliance with applicable regulations and standards been achieved? |
| Efficiency | Behaviour in time | How long does it take to perform a specific task? | How long does it take to perform a specific task? |
| | Resources Uses | How much resources is needed to perform operations (memory,....)? | Is the software capable of performing operations with expected capacity and resources? |
| | Compliance | What is the compliance of system efficiency with applicable regulations and standards? | What is the compliance of system efficiency with applicable regulations and standards? |

Table 2. Cont.

| Criterion | Sub-Criteria | Internal Measure | External Measure |
|---|---|---|---|
| Suitability for maintenance | Possibility of analysis | How good are the functions that provide diagnosis of errors? | Can a user easily identify and correct errors? |
| | Interchange-ability | Is the process of software changes simple? | Can user's problems be adequately resolved in an acceptable time? |
| | Stability | Do certain modifications affect the stability of the software? | Can the software be used without errors after the maintenance process? |
| | Testing | What is the possibility of software testing? | Can a user easily perform operational software testing? |
| | Compliance | What is the compliance of system maintenance with applicable regulations and standards? | What is the compliance of system maintenance with applicable regulations and standards? |
| Transfer ability | Adaptivity | What is the ability to adapt products to organizational and other changes? | Can the software be easily adapted to changes in the environment from the user's point of view? |
| | Installation | What is flexibility of software to be installed in a particular environment? | Can users easily install software in a particular environment? |
| | Coexistence | What is the possibility of connecting this software with other products without a bad interaction? | How often do you make mistakes when connecting with other products? |
| | Replacement | How many functions must remain unchanged in software product? | Does the addition of new elements provide the same functionality from the user's point of view? |
| | Compliance | What is the compliance of the system's portability with applicable regulations and standards? | What is the compliance of the system's portability with applicable regulations and standards? |

Experts performed their task using individual survey sheets, provided by the researchers. In this way, the experts had not idea of others' opinion or final results. The researchers used a commercial software, Expert Choice, for managing and analysing data. In particular, in the case of inconsistency in values, the researches supported the experts in the way to re-examine those aspects which caused the data inconsistency. This procedure was in line with standards.

*5.2. Approach to the Evaluation of Models (Software Solutions) Based on ISO Standards and AHP Method*

Among the standards for evaluation, it is possible to list ISO/IEC 9126 and ISO/IEC 14598, used in the case of developed or ready-made software products. The ISO/IEC 14598: *Software product evaluation* represents a support for applying ISO/IEC 9126. Therefore, standard ISO/IEC 9126 defines general purpose of quality of software solutions, quality criteria and provides examples of measures, while ISOIEC 14598 provides directives in the process of evaluation of software products.

The recommendation of ISO/IEC 14598 standard (later replaced by the less known ISO 25040, with minor changes in evaluation procedure) is that the software evaluation procedure shall include all phases as presented in Figure 6.

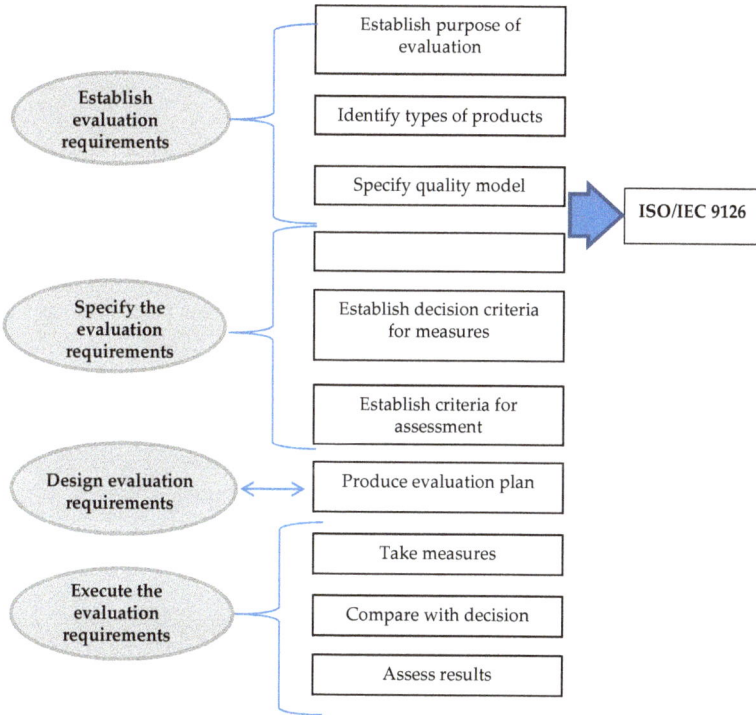

**Figure 6.** Software solutions evaluation procedure in compliance with the ISO/IEC 14598 standard.

Even if this procedure is quite standardized, the related recommendations are not so simple to be implemented: between other aspects, a clear choice of the specific methods to be used is missing.

With the aim of compliance with the evaluation procedure recommended by the ISO/IEC 14598, the AHP method probably represents the best way: [52–58] report, in fact, that AHP can be considered as the most common procedure in the software products evaluation, whereas evaluation criteria are based on the recommendations under standard ISO/IEC 9126.

In Reference [59], in particular, the application of AHP in choosing the ERP software in production companies has been shown. In Reference [60], the authors have created their own evaluation model in the Component Based Development (CBD) process, respecting the standards ISO 9126 and ISO 14598. Such evaluation model has been tested by applying AHP method in order to confirm the model validity showing satisfactory results.

The general recognition regarding the AHP validity is the reason why this method has be used in the study as a tool to provide as objective model evaluation and as much as possible in compliance with the criteria provided by ISO/IEC 9126 and with the procedure which is fully compatible with the evaluation approach defined under standard ISO/IEC 14598.

*5.3. Analytic Hierarchy Process*

In brief, it is a quite common idea that any problem that requires structuring, measurement and synthesis of the results represents a good opportunity for applying AHP.

The AHP method has many advantages as following [54,61–65]:

- the method is based on a well-defined mathematical model;
- quantification of elements structured in a hierarchical model with the possibility of documenting;
- applicable in situations involving a number of criteria;
- allows subjective assessment;
- uses qualitative and quantitative data;
- allows to measure consistency;
- it is widely analysed and applied in scientific literature;
- it has very good software support;
- suitable for group decision making.

Furthermore, during its use, the method offered additional advantages [66]:

- AHP successfully simulates the decision-making process in all its phases: from defining the goals, criteria and alternatives to evaluating criteria and alternatives and obtaining results.
- AHP gives to the decision maker information about the weight coefficients of the criteria relative to the target providing useful index for monitoring the efficacy of the procedure.
- When used in group decision-making, AHP significantly improves communication between group members [67,68], which results in better understanding and finding an easier path to consensus and as a final result provides group members with more confidence in the chosen alternative.
1. AHP successfully identifies problems and points to the inconsistency of decision makers.

*5.4. Models Ranking*

Results achieved with the Team 1 evaluation (experts not employed in the company) using the AHP method in the commercial software program Expert Choice are shown in Figure 7. They rank Model 2 as the best one. Model 2 consists of ECO perspective in the conventional BSC and ECO BSC focused on the Environmental management system. Results achieved by the Team 1 evaluation (experts not employed in the company) using the AHP method in the commercial software program Expert Choice are shown in Figure 7.

In the same way, the application of the AHP method in the commercial software program *Expert Choice* helped during the assessment performed by Team 2 (managers from the company), with final results reported in Figure 8. Also in this case, it can be seen that Model 2 shows an important advantage over all other models. At the same time, Team 2 favoured models which imply creation of a separate ECO BSC system (Models 1 and 2) while Model 3 and 4 are estimated with drastically lower marks.

In order to get the final hierarchy for the recommendation of models, an additional AHP approach is used which merged the results obtained through the assessment performed by both teams (Figure 9).

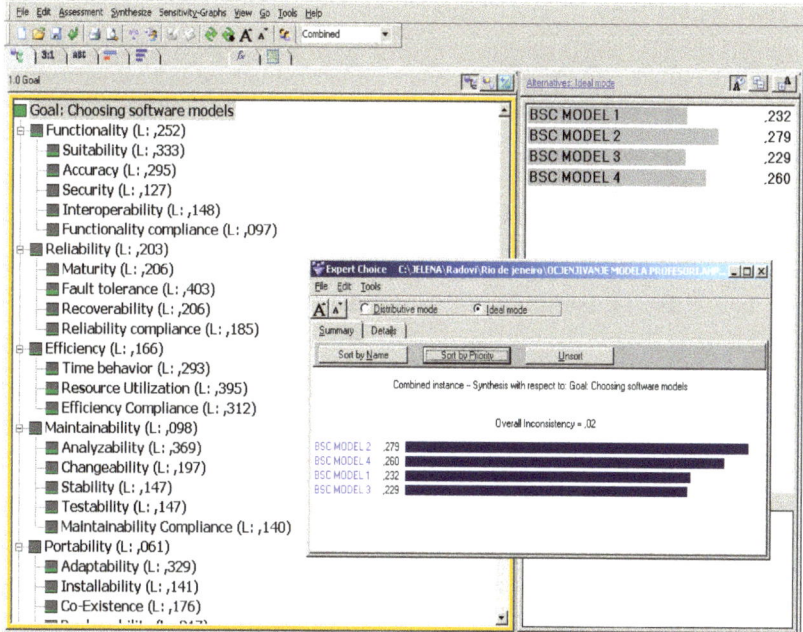

**Figure 7.** Results of evaluation of the BSC models obtained by Team 1 using the AHP method.

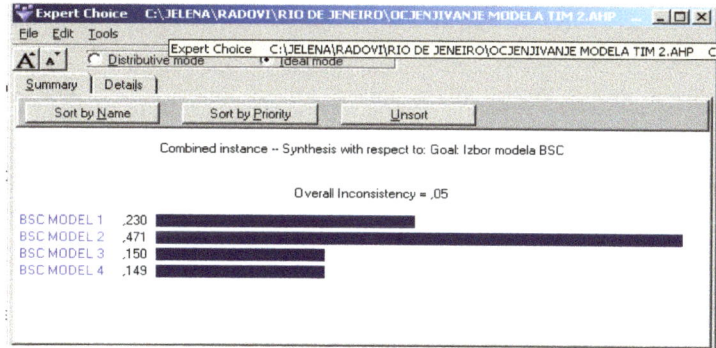

**Figure 8.** Results of evaluation of the BSC models obtained by Team 2 using the AHP method.

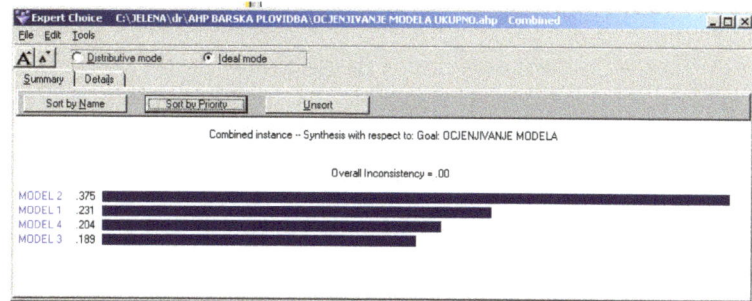

**Figure 9.** Final results of ranking of the models.

## 5.5. Further Details on Models

Through this assessment, supported in method by the application of the AHP approach in the evaluation procedure we can notice a certain rule of ranking the models related to the volume of their orientation to the environmental management system. Following considerations were made:

- The best estimates were given to the models that most broadly encompass the EMS (Model 1 and 2);
- Final evaluation result has shown that Model 2, which involves ECO BSC and Eco perspective in conventional BSC, is the best ranked one. This model was also the best evaluated one by each team individually, what was not the case with Model 1.
- The third ranked model is Model 4 which includes eco perspective in conventional model but does not have ECO BSC. This model has stronger orientation towards the environmental protection than Model 3, since Eco perspective in conventional BSC offers more space for environmental metrics.
- The lowest ranked model in final evaluation result is Model 3, although it is the most frequently applied one. The reason for this shall be sought in the fact that organisations that have implemented BSC are mainly profit oriented organisations which are not strategically oriented to environmental protection.

As main methodological outcome, these results confirm the risky hypothesis, which argues that the BSC system with the specially created ECO BSC, based on the concept of a non-profit organization and focused solely to the environmental management system, produces an efficient management system that improves ecological performance.

Finally, Figure 10 summarizes the main results dealing with the models' assessment. In particular, it shows by lines the capability of each model to reach the specific goal and in histograms the alternative options they take into account.

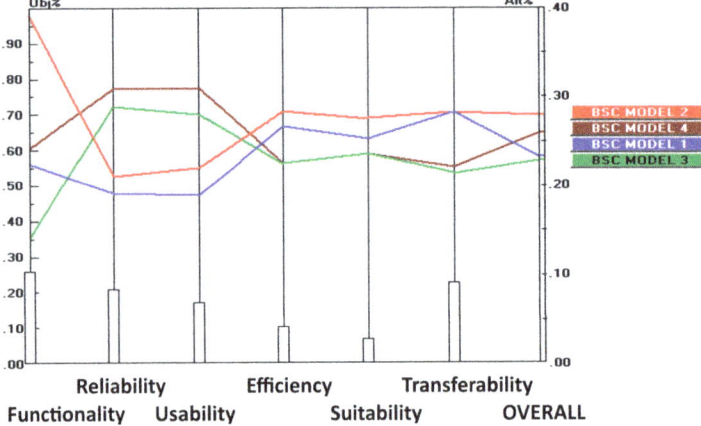

**Figure 10.** Details on comparing models (obj% = objective goal, in lines; al% = alternative options, in histograms).

## 5.6. Comparative Analysis

Tables 3 and 4 show comparison between the 4 BSC models, both in terms of structure and consistency.

**Table 3.** Details in the comparative analysis of the 4 recommended BSC models.

| Model Characteristics | Model 1 | | Model 2 | | Model 3 | Model 4 |
|---|---|---|---|---|---|---|
| Existence of the ECO BSC | Y | | Y | | N | N |
| Existence of a Budget perspective | Y | | Y | | N | N |
| Additional ECO Perspective | N | | Y | | N | Y |
| Profit Organization | N | Y | N | Y | Y | Y |
| Base of the strategic map | ECO | BSC | ECO | BSC | BSC | BSC |
| Inclusion of the significance of goals in the perspectives obtained with the application of AHP | IP ST | US LD | IP ST LD | LD ECO | No application of significance coefficients in the goals | No application of significance coefficients in the goals |
| Inclusion of coefficients of significance of measures in AHP targets | IP (AG) ST LD (AG) | IP (CG) US LD (CG) | IP (AG) ST (AG) LD (AG) | LD ECO (AG) | No application of coefficients of significance of measures in the objectives | No application of coefficients of significance of measures in the objectives |

Legend: Yes (**Y**); No (**N**); Internal Processes (**IP**); Learning and Development (**LD**); Stakeholders (**ST**); Users (**US**); All Goals (**AG**); Company goals (**CG**).

**Table 4.** Numbers of targets, measures and objectives in the 4 recommended BSC models.

| Model | Model 1 | | | | | | | | Model 2 | | | | | | | |
|---|---|---|---|---|---|---|---|---|---|---|---|---|---|---|---|---|
| | ECO | | | | BSC | | | | ECO | | | | BSC | | | |
| Number of | A | B | C | D | A | B | C | D | A | B | C | D | A | B | C | D |
| Eco targets in the scorecard | 10 | 10 | 10 | 10 | 3–13 | 3–9 | 3–10 | 3–10 | 10 | 10 | 10 | 10 | 5–12 | 5–11 | 5–12 | 5–12 |
| Eco measures in the scorecard | 33 | 34 | 21 | 29 | 7–27 | 8–21 | 5–15 | 8–18 | 33 | 34 | 21 | 29 | 4–24 | 5–18 | 2–12 | 5–15 |
| Related objectives | 9 | 6 | 6 | 6 | 6 | 6 | 6 | 6 | 8 | 8 | 8 | 8 | 8 | 8 | 8 | 8 |
| Associated measures | 9 | 9 | 8 | 8 | 6 | 6 | 8 | 8 | 6 | 6 | 5 | 5 | 3 | 3 | 5 | 5 |

| Model | Model 3 | | | | Model 4 | | | |
|---|---|---|---|---|---|---|---|---|
| | BSC | | | | BSC | | | |
| Number of | A | B | C | D | A | B | C | D |
| Eco targets in the scorecard | 3–13 | 3–9 | 3–10 | 3–10 | 5–12 | 5–12 | 5–12 | 5–12 |
| Eco measures in the scorecard | 10–32 | 11–24 | 6–16 | 11–21 | 19–38 | 20–33 | 10–20 | 17–26 |
| Related objectives | 5 | 5 | 5 | 5 | 7 | 7 | 7 | 7 |
| Associated measures | 12 | 11 | 7 | 11 | 21 | 20 | 11 | 16 |

Legend: Scorecard at the corporate level (**A**); Scorecard at Ship level (**B**); Scorecard at the Travel Agency level (**C**); Scorecard at the Catering level (**D**).

In particular, Table 3 summarizes the inclusion/exclusion of determined perspectives (e.g., budget, eco-sustainability) in the analysis in respect to each model. It can be noted, for instance, that a strategic map based on ECO targets could be considered only for Models 1 and 2.

A clear indication of the profit/non-profit orientation is also reported, together with the applicability of specific coefficients in AHP targeting. Amongst other potential approaches, the investigation was focused on evaluating the effect of changes in the goals between the following options: Internal Processes (IP); Learning and Development (LD); Stakeholders (ST) or Users (US).

Bearing in mind these options, the effects of changing the goals when choosing between a more general attention of All Goals (AG) and the most relevant goals (CG) for the company, such as the profit were also considered.

The following text contains a more detailed analysis of the results.

Table 4 details the values of targets, measures, objectives adopted during the analysis. In particular, it reports the number of:

- Eco targets in the scorecard (in relation to the total number of goals)
- Eco measures in the scorecard (in relation to the total number of measures)
- Related objectives (of the conventional BSC with eco goals at the same level)
- Associated measures (of conventional BSC with eco measures)

In particular, four different approaches were considered for defining the scorecards: Scorecard at the corporate level (A); Scorecard at Ship level (B); Scorecard at the Travel Agency level (C); Scorecard at the Catering level (D). Also in this case, it can be seen that a deeper analysis (in terms of combinations) has been performed for Model 1 and 2.

When determining the number of measures and targets (in Table 4), measures that belong to the second hierarchy level were not included because their values were included on a higher level of measures by simple summation. It should be also noted that Models 1 and 3 created a specific objective, which incorporates 3 sub-objectives (controlled aspects, potential hazards and regulatory compliance) for which there are no explicitly defined measures and these sub-targets were evaluated as measures in the table. Analysing the Tables, it can be noted that there are considerable differences between the models. Namely, Models 1 and 2 contain ECO BSC and therefore have a lower number of eco targets and measures explicitly included in the conventional BSC model, while Models 3 and 4 directly include selected eco targets and measures.

In this way, the total number of objectives and measures of Models 3 and 4 increases significantly, with their connection entirely based on models of profit organizations that would correspond to the concept of sustainable development that is not fully justified and applicable to the operations of the company.

Also, in the process of building the ECO BSC model, the significance of metrics was evaluated using the group AHP method to determine the key points that can be operated in order to improve the business results. The results of such measurements were included only in Models 1 and 2 at the corporate level of the ECO BSC and then transferred to the level of the conventional BSC model to the extent to which it ensured feasible connecting.

*5.7. Applicability*

The synthesis of the evaluation results has shown that those models which include a separately created ECO BSC have a better ranking (in terms of all previously considered criteria) with respect to those models only comprising of a certain number of goals and measures within the conventional BSC model. This general result is exactly where the major contribution of this paper lies. The same model ranking for strategically oriented to environmental protection has been obtained in consideration of a specific real situation and involve a well-definite for-profit organisation. At the same time, it implies that the same recommendations are almost valid in respect to those organisations having similar dilemma with regard to the choice of management's approach to the performance based on a BSC approach. In details, the recommendations are:

- Organisations strategically oriented to environmental protection shall choose Model 1 or Model 2 since these models contain separate ECO BSC which completely covers the entire environmental protection field of an organisation. These models provide a good connection between the ECO BSC and a conventional BSC model, therefore avoiding the possibility of creating parallel systems on one side and on the other side, providing for the conventional BSC model not to be over-loaded with environmental protection goals and measures, what would make environmental protection a priority over other issues.
- Organisations not strategically oriented to environmental protection or those that have less significant impact on environment, may choose Model 3 or Model 4, depending on their commitment to this issue. Model 4 is more oriented to the environmental protection due to a special eco perspective, while Model 3 contains only some of the environmental protection targets and measures within existing perspectives.
- Furthermore, organisations aiming to implement the ISO 14001 standard but not in a position to implement BSC system on the level of the entire organisation, shall establish the network of goals and measures following the model ECO BSC, so as to enable more efficient monitoring, measuring and improvement of ecological performance, what has been assessed as the greatest shortcoming in application of ISO 14001 standard. ECO BSC model has been created on the operating principle of non-profit organisations, with the aim to grow into a profitable model in case that the amounts of budget and profit from environmental protection management come even in justified conditions. Such approach is also a logical one with regard to environmental protection management, much

more logical that the one based on the operating principle of for-profit organisations which includes provision of financial benefits form environmental protection.

An additional contribution of the paper may also be found in the manner of creating the ECO BSC model. Explicitly, ECO BSC has been made on the operating principle of non-profit organisations with the budget as a basic perspective. Thus, the goals set in the budget perspective shall lead to the fulfilment of other perspectives, what is the approach which is suitable for an objective approach to the environmental protection management in most organisations.

Although created on the operating principle of non-profit organisations, having budget as a starting perspective and finance perspective on the top of a strategic map, ECO BSC still tends to grow into a profitable model if the amounts of budget and profit from environmental protection management come even in justified conditions.

## 6. Conclusions

This manuscript focuses on environmental sustainability issue, specifically on environmental management systems (EMS) and its reading and application by combining the EMS and BSC approach.

This methodological study also explores the case of one marine organization in Montenegro, used as a case study for testing this integrated application of the EMS and the BSC.

The field of investigation receives still little attention by researchers and several suggestions are provided in terms of integrated and more effective managerial instruments and practices that can be adopted to face the barriers that still exist in applying and following the managerial approach based on the respect for the environmental.

In particular, in this paper, four models of the BSC have been presented which, in different ways, include the metric of the EMS applied to the relevant case of the company operating in the marine transport field. Models 1 and 2 include the so-called ECO BSC component, created in relation to the concept of non-profit organizations. Models 3 and 4 do not contain this ECO BSC component, oriented to environmental protection. In that case, this aspect is covered by the framework of already existing perspectives and/or additionally created perspective in the conventional BSC.

Expert assessment, based on the AHP method and with respect to the criteria stipulated under the ISO 9126 standard, largely ranked Model 2 as the best one. This method is based on the special criteria of the ECO BSC system that covers the area of the EMS and has ECO perspective in the conventional BSC system that is able to connect these two systems.

This kind of model considers organization strategically oriented to the environmental protection in line with the robust requirements stipulated by international standards and laws on marine transport organizations in the field of environmental protection.

This general result should be a recommendation for all organizations committed to the EMS which have dilemmas when opting for the BSC approach with EMS elements.

As specific results, the advantages on applying the models 1 and 2 can be listed, in brief, as a:

1. continuous monitoring and measurement of all ecological performance, which proved to be the biggest drawback of the implementation of the ISO 14001 standards;
2. comprehensive knowledge about the relationship between ecological performance and strategic goals;
3. comprehensive knowledge about the relationship between ecological performance and user perspective/user satisfaction;
4. comprehensive knowledge about the relationship between EMS and Finance in order to measure the ecological efficiency (if this can be said);
5. database containing information to be considered for more efficient decision making in the field of Environmental Management strategies.

Finally, beyond any specific results of the present case, this qualitative investigation also provides information and compares different alternative methods to connect the ECO and the conventional BSC system avoiding the creation of parallel management systems, in a single ECO BSC system able to measures ecological performance in a new environmental management system.

Regarding the future, this analysis could be conveniently oriented to different directions. From one side, the current study can be validated in (at least) a couple of years. In fact, after a reasonable period of time, a consistent database will be available that will allow discussion on assumptions and predictions.

On the other hand, it may be interesting to repeat the same analysis taking in consideration a different maritime transport company. It would be ideal to define a new case study that would present different conditions, while preserving correspondence in a way that simplifies the analysis. For example, it would be extremely interesting to involve a company active in the passenger transport between islands of Croatia or Greece. In this way, general aspects such as variations in the tourist demand for services but also local ones, such as fuel and personnel costs, could be compared.

**Author Contributions:** Conceptualization, J.S.J. and C.F.; methodology, J.S.J. and Z.K.; software, A.V.; validation, J.S.J., C.F. and Z.K.; formal analysis, A.V.; investigation, J.S.J.; data curation, C.F.; writing—original draft preparation, J.S.J.; writing—review and editing, C.F.; visualization, Z.K. and A.V.; supervision, Z.K.; project administration, A.V.; funding acquisition, A.V.

**Funding:** This research received no external funding.

**Acknowledgments:** Special thanks to Ana Pavlovic for her support during the phases of writing and editing.

**Conflicts of Interest:** The authors declare no conflict of interest. The company, selected as case-study, had no role in the design of the study; in the analyses or interpretation of data; in the writing of the manuscript or in the decision to publish the results.

## References

1. Gomez, A.; Rodriguez, M.A. The effect of ISO 14001 certification on toxic emissions: An analysis of industrial facilities in the north of Spain. *J. Clean. Prod.* **2011**, *19*, 1091–1095. [CrossRef]
2. Franchetti, M. ISO 14001 and solid waste generation rates in US manufacturing organizations: An analysis of relationship. *J. Clean. Prod.* **2011**, *19*, 1104–1109. [CrossRef]
3. Naveh, E.; Marcus, A.A. When does the ISO 9000 quality assurance standard lead to performance improvement? Assimilation and going beyond. *IEEE Trans. Eng. Manag.* **2004**, *51*, 352–363. [CrossRef]
4. Potoski, M.; Prakash, A. Covenants with weak swords: ISO 14001 and facilities' environmental performance. *J. Policy Anal. Manag.* **2005**, *24*, 745–769. [CrossRef]
5. Reinhardt, F. Market failure and the environmental policies of firms. *J. Ind. Ecol.* **1997**, *3*, 9–21. [CrossRef]
6. Matthews, D.H. Assessment and Design of Industrial Environmental Management Systems. Ph.D. Thesis, Carnegie-Mellon University, Pittsburgh, PA, USA, 2001.
7. Hamschmidt, J.; Dyllick, T. ISO 14001: Profitable—Yes! But is it Eco—Effective? *Greener Manag. Int.* **2001**, *34*, 43–54. [CrossRef]
8. Dahlstrom, K.; Howes, C.; Leinster, O.; Skea, J. Environmental management systems and company performance: Assessing the case for extending risk-based regulation. *Eur. Environ.* **2003**, *13*, 187–203. [CrossRef]
9. Scalet, R.; Yokohama, A.; Koscianski, A.; Rêgo, C.M.; Asanome, C.; Romero, D.; Vostoupal, T.M. ISO/IEC 9126 and 14598 integration aspects: A Brazilian viewpoint. In Proceedings of the Second World Congress on Software Quality, Yokohama, Japan, 25 September 2000; p. 350.
10. Helms, M.M.; Nixon, J. Exploring SWOT analysis—Where are we now?: A review of academic research from the last decade. *J. Strategy Manag.* **2010**, *3*, 215–251. [CrossRef]
11. Ifediora, C.O.; Idoko, O.R.; Nzekwe, J. Organization's stability and productivity: The role of SWOT analysis an acronym for strength, weakness, opportunities and threat. *Int. J. Innov. Appl. Res.* **2014**, *2*, 23–32.
12. Madhani, M.P. Marketing and Supply Chain Management Integration: A Resource-Based View of Competitive Advantages. *Int. J. Value Chain Manag.* **2012**, *6*. [CrossRef]
13. Mladenović, S.; Milosavljević, P.; Milojević, N.; Pavlović, D.; Todorović, M. The path towards achieving a lean six sigma company using the example of the Shinwon company in Serbia. *Facta Univ. Ser. Mech. Eng.* **2016**, *14*, 219–226. [CrossRef]

14. Fragassa, C.; Pavlovic, A.; Massimo, S. Using a Total Quality Strategy in a new Practical Approach for Improving the Product Reliability in Automotive Industry. *Int. J. Qual. Res.* **2014**, *8*, 297–310.
15. Fragassa, C. From Design to Production: An integrated advanced methodology to speed up the industrialization of wooden boats. *J. Ship Prod. Des.* **2017**, *33*, 237–246. [CrossRef]
16. Cano, J.A.; Vergara, J.J.; Puerta, F.A. Design and implementation of a balanced scorecard in a Colombian company. *Rev. Espac.* **2017**, *38*, 19. Available online: http://www.revistaespacios.com/a17v38n31/17383119.html (accessed on 25 April 2019).
17. Jovanović, J. Model of Improving Environmental Management System Using Multisoftware. Ph.D. Thesis, Faculty of Mechanical Engineering, Podgorica, Montenegro, 2009.
18. Kaplan, R.S.; Norton, D.P. *The Strategy-Focused Organization: How Balanced Scorecard Companies Thrive in the New Business*; Harvard Business School: Boston, MA, USA, 2001.
19. Karasneh, A.A.; Al-Dahir, A. Impact of IT-Balanced Scorecard on Financial Performance: An Empirical Study on Jordanian Banks. *Eur. J. Econ. Financ. Adm. Sci.* **2012**, *46*, 54–70.
20. Niven, P.R. *BSC Step by Step for Government and Non Profit Agencies*; John Wiley and Sons: Hoboken, NJ, USA, 2003.
21. Dvorski, D. Pokazatelji uspješnosti poslovanja primjenom modela uravnoteženih ciljeva (Indicators of Business Performance by Applying the Model of Balanced Goals). Master's Thesis, University of Zagreb, Zagreb, Croatia, 2005.
22. Zwyalif, I.M. Using a Balanced Scorecard Approach to Measure Environmental Performance: A Proposed Model. *Int. J. Econ. Financ.* **2017**, *9*. [CrossRef]
23. Kaplan, R.S.; Norton, D.P. *Alignement Using the BSC to Create Corporate Synergies*; Harvard Business School: Harvard, UK, 2006.
24. Kaplan, R.; Norton, D. *Putting the Balanced Scorecard—Translating Strategy into Action*; Harvard Business School Press: Harvard, UK, 2001.
25. Figge, F.; Hahn, T.; Schaltegger, S.; Wagner, M. The Sustainability Balanced Scorecard—Theory and Application of a Tool for Value-Based Sustainability Management. In Proceedings of the Greening of Industry Network Conference, Gothenburg, Corporate Social Responsibility-Governance for Sustainability, Göteborg, Sweden, 23–26 June 2002.
26. Butler, J.B.; Henderson, S.C.; Raiborn, C. Sustainability and the balanced scorecard: Integrating green measures into business reporting. *Manag. Account. Q.* **2011**, *12*, 1–10.
27. Van der Woerd, F.; Brink, T.W.M. Feasibility of a responsive business scorecard—A pilot study. *J. Bus. Ethics* **2004**, *55*, 173–186. [CrossRef]
28. Bento, R.F.; Mertins, L.; White, L.F. Ideology and the balanced scorecard: An empirical exploration of the tension between shareholder value maximization and corporate social responsibility. *J. Bus. Ethics* **2016**, *142*, 769–789. [CrossRef]
29. Bieker, T.; Gminder, C.U. *Towards a Sustainability Balanced Scorecard. Environmental Management & Policy and Related Aspects of Sustainability*; University of St. Gallen: Gallen, Switzerland, 2001. Available online: http://www.oikos-international.org/fileadmin/oikosinternational/international/Summer_Academies_old_ones_/edition_2001/Papers/Paper_Bieker_Gminder.pdf (accessed on 25 April 2019).
30. Epstein, M.J.; Wisner, P.S. Using a Balanced Scorecard to implement Sustainability. *Environ. Qual. Manag.* **2001**, *11*, 1–10. [CrossRef]
31. Bieker, T.; Waxenberger, B. Sustainability Balanced Scorecard and business ethics—Using the BSC for Integrity Management. In Proceedings of the 10th International Conference of the Greening of Industry Network, Göteborg, Sweden, 23–26 June 2002.
32. Olve, N.G.; Roy, J.; Wetter, M. *Performance Drivers, A Practical Guide to Using the Balanced Scorecard*; John Wiley and Sons: New York, NY, USA, 2004.
33. Sidiropoulos, M.; Mouzakitis, Y.; Adamides, E.; Goutsos, S. Applying Sustainable Indicators to Corporate Strategy: The Eco-Balanced Scorecard. *Environ. Res. Eng. Manag.* **2004**, *1*, 28–33.
34. Gminder, C.U. Environmental management with the Balanced Scorecard. A case study of the Berlin Water Company, Germany. In *The Business of Water and Sustainable Development*; Greenleaf Publishing: Sheffield, UK, 2005; pp. 51–62.
35. Bressloff, P.C.; Weir, D.J. Neural Networks. *Gec J. Res.* **1991**, *8*, 151–169.

36. Gao, T. *Lives in the Balance: Managing with the Scorecard in Not-for-Profit Healthcare Settings*; Jinan Central Hospital: Jinan, China; University of South Australia: Adelaide, Australia, 2006.
37. Hansen, E.G.; Schaltegger, S. *Pursuing Sustainability with the Balanced Scorecard: Between Shareholder Value and Multiple Goal Optimization*; Centre for Sustainability Management: Lüneburg, Germany, 2012.
38. Krivokapić, Z.; Jovanović, J. Using Balanced Scorecard to improve Environmental management system. *Stroj. Vestn.* **2009**, *55*, 262–279.
39. Jovanovic, J. Management of the organization based on balanced scorecards. *Int. J. Qual. Res.* **2011**, *5*, 317–325.
40. Bieker, T. Sustainability management with the Balanced Scorecard. In Proceedings of the 5th International Summer Academy on Technology Studies, Deutschlandsberg, Austria, 13–19 July 2003; pp. 17–34.
41. Zingales, F.; Hockerts, K. *Balanced Scorecard and Sustainability: Examples from Literature and Practices*; Working Paper 30; INSEAD: Fontainebleau, France, 2003; Available online: http://flora.insead.edu/fichiersti_wp/inseadwp2003/2003-30.pdf (accessed on 25 April 2019).
42. Hahn, T.; Figge, F. Why architecture does not matter: On the fallacy of sustainability balanced scorecards. *J. Bus. Ethics* **2018**, *150*, 919–935. [CrossRef]
43. Hansen, E.G.; Schaltegger, S. Sustainability balanced Scorecards and their Architectures: Irrelevant or Misunderstood? *J. Bus. Ethics* **2018**, *150*, 937–952. [CrossRef]
44. Di Vaio, A.; Varriale, L. Management innovation for environmental sustainability in seaports: Managerial accounting instruments and training for competitive green ports beyond the regulations. *Sustainability* **2018**, *10*, 783. [CrossRef]
45. Di Vaio, A.; Varriale, L.; Alvino, F. Key performance indicators for developing environmentally sustainable and energy efficient ports: Evidence from Italy. *Energy Policy* **2018**, *122*, 229–240. [CrossRef]
46. Sislian, L.; Jaegler, A. A sustainable maritime balanced scorecard applied to the Egyptian Port of Alexandria. *Supply Chain Forum Int. J.* **2018**, *19*, 101–110. [CrossRef]
47. Möller, A.; Schaltegger, S. The Sustainability Balanced Scorecard as a Framework for Eco-Efficiency Analysis. *J. Ind. Ecol.* **2005**, *9*, 73–83. [CrossRef]
48. Schaltegger, S.; Wagner, M. Integrative management of sustainability performance, measurement and reporting. *Int. J. Acc. Audit. Perform. Eval.* **2006**, *3*, 1–19. [CrossRef]
49. Bieker, T.; Dyllick, T.; Gminder, C.U.; Hockerts, K. *Towards a Sustainability Balanced Scorecard Linking Environmental and Social Sustainability to Business Strategy*; Institute for Economy and the Environment: Tigerbergstrasse, Switzerland, 2001.
50. Scavone, G.M. Challenges in Internal Environmental Management Reporting in Argentina. *J. Clean. Prod.* **2006**, *14*, 1276–1285. [CrossRef]
51. Johnson, D.S. Identification and selection of environmental performance indicators: Application of the Balanced Scorecard approach. *Corp. Environ. Strategy* **1998**, *5*, 34–41. [CrossRef]
52. Schniederjans, M.J.; Wilson, R.L. Using the analytic hierarchy process and goal programming for information system project selection. *Inf. Manag.* **1991**, *20*, 333–342. [CrossRef]
53. Wei, C.C.; Dhien, C.F.; Wang, M.J. An AHP-based approach to ERP system selection. *Int. J. Prod. Econ.* **2005**, *96*, 47–62. [CrossRef]
54. Lien, C.T.; Chan, H.L. A Selection Model for ERP System by Applying Fuzzy AHP approach. *Int. J. Comput. Internet Manag.* **2007**, *15*, 58–72.
55. Peng, Y.; Kong, G.; Wang, G.; Wu, W.; Shi, Y. Ensemble of software defect predictors: An AHP-based evaluation method. *Int. J. Inf. Technol. Decis. Mak.* **2011**, *10*, 187–206. [CrossRef]
56. Zhu, L.; Aurum, A.; Gorton, I.; Jeffery, R. Tradeoff and Sensitivity Analysis in Software Architecture Evaluation Using Analytic Hierarchy Process. *Softw. Qual. J.* **2005**, *13*, 357–375. [CrossRef]
57. Kanellopoulos1, Y.; Antonellis, P.; Antoniou, D.; Makris, C.; Theodoridis, E.; Tjortjis, C.; Tsirakis, N. Code quality evaluation methodology using the ISO/IEC 9126 Standard. *Int. J. Softw. Eng. Appl.* **2010**, *1*, 17–36. [CrossRef]
58. Al-Naeem, T.; Gorton, I.; Babar, M.A.; Rabhi, F.; Benatallah, B. A quality-driven systematic approach for architecting distributed software applications. In Proceedings of the 27th International Conference on Software Engineering (ICSE), St. Louis, MO, USA, 15–21 May 2005.
59. Perera, H.S.C.; Costa, W.K.R. Analytic Hierarchy Process for Selection of Erp Software for Manufacturing Companies. *Vis. J. Bus. Perspect.* **2008**, *12*, 1–11. [CrossRef]

60. Lee, K.; Lee, S.J. A Quantitative Evaluation Model Using the ISO/IEC 9126 Quality Model in the Component Based Development Process. In Proceedings of the International Conference on Computational Science and Its Applications, Glasgow, UK, 8–11 May 2006.
61. Multi-Criteria_Decision_Analysis. Available online: http://en.wikipedia.org/wiki/Multi-criteria_decision_analysis/ (accessed on 25 April 2019).
62. Mogharreban, N. Adaptation of Cluster Discovery Technique to a Decision Support System. *Interdiscip. J. Inf. Knowl. Manag.* **2006**, *1*, 59–69. [CrossRef]
63. Xu, L.; Yang, J.-B. *Introduction to Multi-Criteria Decision Making and the Evidential Reasoning Approach*; Manchester School of Management: Manchester, UK, 2001; pp. 1–106.
64. Tahriri, F.; Osman, M.R.; Ali, A.; Yusuff, R.M. A Review of Supplier Selection Methods in Manufacturing Industries. *Suranaree J. Sci. Technol.* **2008**, *15*, 201–208.
65. Selecting the Ideal FPGA Vendor for Military Programs, White Paper. Available online: http://www.altera.com/literature/wp/wp-01094-select-military-vendor.pdf (accessed on 25 April 2019).
66. Jandrić, Z.; Srđević, B. Analytical hierarchical process to support decision-making in water management ('Analitički hijerarhijski proces kao podrška donošenju odluka u vodoprivredi'). *Vodoprivreda* **2000**, *32*, 324–334.
67. Karlsson, J.; Wohlin, C.; Regnell, B. An evaluation of methods for prioritizing software requirements. *Inf. Softw. Technol.* **1998**, *39*, 939–947. [CrossRef]
68. Harker, P.T.; Vargas, L.G. The theory of ratio scale estimation: Saaty Analytic hierarchy process. *Manag. Sci.* **1987**, *33*, 1383–1403. [CrossRef]

© 2019 by the authors. Licensee MDPI, Basel, Switzerland. This article is an open access article distributed under the terms and conditions of the Creative Commons Attribution (CC BY) license (http://creativecommons.org/licenses/by/4.0/).

MDPI  
St. Alban-Anlage 66  
4052 Basel  
Switzerland  
Tel. +41 61 683 77 34  
Fax +41 61 302 89 18  
www.mdpi.com

*Journal of Marine Science and Engineering* Editorial Office  
E-mail: jmse@mdpi.com  
www.mdpi.com/journal/jmse

www.ingramcontent.com/pod-product-compliance
Lightning Source LLC
LaVergne TN
LVHW071953080526
838202LV00064B/6733